Naturally Selected

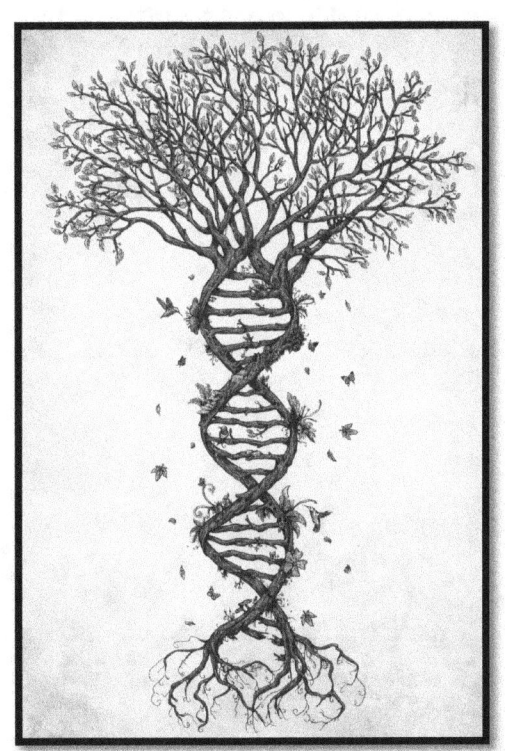

ORIGINAL READINGS IN
EVOLUTIONARY BIOLOGY

Format Copyright © 2016 by MSAC Philosophy Group

All rights reserved. This book or any portion thereof may not be reproduced or used in any manner whatsoever without the express written permission of the publisher except for the use of brief quotations in a book review or scholarly journal.

First Edition: 2016

ISBN: 978-1-56543-700-5

MSAC Philosophy Group, Mt. San Antonio College

1100 Walnut, California 91789 USA

Website: http://www.neuralsurfer.com

Imprint: The Runnebohm Library Series

Editor: David Christopher Lane

Naturally Selected

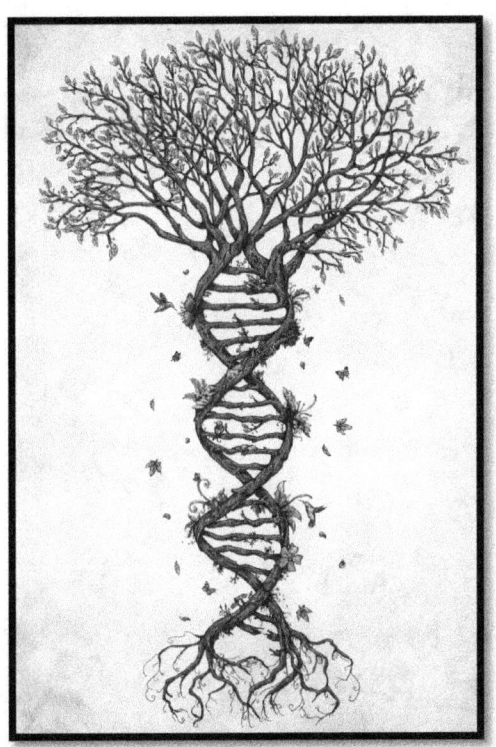

ORIGINAL READINGS IN
EVOLUTIONARY BIOLOGY

2016
Mt. San Antonio College
Walnut, California

Format Copyright © 2016 by MSAC Philosophy Group

All rights reserved. This book or any portion thereof may not be reproduced or used in any manner whatsoever without the express written permission of the publisher except for the use of brief quotations in a book review or scholarly journal.

First Edition: 2016

ISBN: 978-1-56543-700-5

MSAC Philosophy Group, Mt. San Antonio College

1100 Walnut, California 91789 USA

Website: http://www.neuralsurfer.com

Imprint: The Runnebohm Library Series

Editor: David Christopher Lane

TABLE OF CONTENTS

On the Tendency of Varieties to Depart Indefinitely from the Original Type
Alfred Russel Wallace

1

Experiments in Plant Hybridization
Gregor Mendel

9

On the Origin of Species, a Review
Thomas H. Huxley

17

Molecular Structure of Nucleic Acids
James Watson and Francis Crick

45

Naturally Selected

Original Papers in Biology

Issue One

On the Tendency of Varieties to Depart Indefinitely from the Original Type

Alfred Russel Wallace

One of the strongest arguments which have been adduced to prove the original and permanent distinctness of species is, that *varieties* produced in a state of domesticity are more or less unstable, and often have a tendency, if left to themselves, to return to the normal form of the parent species; and this instability is considered to be a distinctive peculiarity of all varieties, even of those occurring among wild animals in a state of nature, and to constitute a provision for preserving unchanged the originally created distinct species.

In the absence or scarcity of facts and observations as to *varieties* occurring among wild animals, this argument has had great weight with naturalists, and has led to a very general and somewhat prejudiced belief in the stability of species. Equally general, however, is the belief in what are called "permanent or true varieties,"--races of animals which continually propagate their like, but which differ so slightly (although constantly) from some other race, that the one is considered to be a *variety* of the other. Which is the *variety* and which the original *species*, there is generally no means of determining, except in those rare cases in which the one race has been known to produce an offspring unlike itself and resembling the other. This, however, would seem quite incompatible with the "permanent invariability of species," but the difficulty is overcome by assuming that such varieties have strict limits, and can never again vary further from the original type, although they may return to it, which, from the analogy of the domesticated animals, is considered to be highly probable, if not certainly proved

It will be observed that this argument rests entirely on the assumption, that *varieties* occurring in a state of nature are in all respects analogous to or even identical with those of domestic animals, and are governed by the same laws as regards their permanence or further variation. But it is the object of the present paper to show that this assumption is altogether false, that there is a general principle in nature which will cause many *varieties* to survive the parent species, and to give rise to successive variations departing further and further from the original type, and which also produces, in domesticated animals, the tendency of varieties to return to the parent form.

Editor Charles H. Smith's Note: This is the famous "Ternate essay" introducing natural selection that Wallace sent to Charles Darwin in February of 1858. This paper, along with excerpts from two unpublished writings by Darwin, was read before a special meeting of the Linnean Society of London on 1 July 1858, and published *on pages 53-62 of Volume 3 of that Society's Proceedings series.*

The life of wild animals is a struggle for existence. The full exertion of all their faculties and all their energies is required to preserve their own existence and provide for that of their infant offspring. The possibility of procuring food during the least favourable seasons, and of escaping the attacks of their most dangerous enemies, are the primary conditions which determine the existence both of individuals and of entire species. These conditions will also determine the population of a species; and by a careful consideration of all the circumstances we may be enabled to comprehend, and in some degree to explain, what at first sight appears so inexplicable--the excessive abundance of some species, while others closely allied to them are very rare.

The general proportion that must obtain between certain groups of animals is readily seen. Large animals cannot be so abundant as small ones; the carnivora must be less numerous than the herbivora; eagles and lions can never be so plentiful as pigeons and antelopes; the wild asses of the Tartarian deserts cannot equal in numbers the horses of the more luxuriant prairies and pampas of America. The greater or less fecundity of an animal is often considered to be one of the chief causes of its abundance or scarcity; but a consideration of the facts will show us that it really has little or nothing to do with the matter. Even the least prolific of animals would increase rapidly if unchecked, whereas it is evident that the animal population of the globe must be stationary, or perhaps, through the influence of man, decreasing. Fluctuations there may be; but permanent increase, except in restricted localities, is almost impossible. For example, our own observation must convince us that birds do not go on increasing every year in a geometrical ratio, as they would do, were there not some powerful check to their natural increase. Very few birds produce less than two young ones each year, while many have six, eight, or ten; four will certainly be below the average; and if we suppose that each pair produce young only four times in their life, that will also be below the average, supposing them not to die either by violence or want of food. Yet at this rate how tremendous would be the increase in a few years from a single pair! A simple calculation will show that in fifteen years each pair of birds would have increased to nearly ten millions!² whereas we have no reason to believe that the number of the birds of any country increases at all in fifteen or in one hundred and fifty years. With such powers of increase the population must have reached its limits, and have become stationary, in a very few years after the origin of each species. It is evident, therefore, that each year an immense number of birds must perish--as many in fact as are born; and as on the lowest calculation the progeny are each year twice as numerous as their parents, it follows that, whatever be the average number of individuals existing in any given country, *twice that number must perish annually*,--a striking result, but one which seems at least highly probable, and is perhaps under rather than over the truth. It would therefore appear that, as far as the continuance of the species and the keeping up the average number of individuals are concerned, large broods are superfluous. On the average all above *one* become food for hawks and kites, wild cats and weasels, or perish of cold and hunger as winter comes on. This is strikingly proved by the case of particular species; for we find that their abundance in individuals bears no relation whatever to their fertility in producing offspring. Perhaps the most remarkable instance of an immense bird population is that of the passenger pigeon of the United States, which lays only one, or at most two eggs, and is said to rear generally but one young one. Why is this bird so extraordinarily abundant, while others producing two or three times as many young are much less plentiful? The explanation is not difficult. The food most congenial to this species, and on which it thrives best, is abundantly distributed over a very extensive region, offering such differences of soil and climate, that in one part or another of the area the supply never fails. The bird is capable of a very rapid and long-continued flight, so that it can pass without fatigue over the whole of the district it inhabits, and as soon as the supply of food begins to fail in one place is able to discover a fresh feeding-ground. This example strikingly shows us that the procuring a constant supply [[p. 56]] of wholesome food is almost the sole condition requisite for ensuring the rapid increase of a given species, since neither the limited fecundity, nor the unrestrained attacks of birds of prey and of man are here sufficient to check it. In no other birds are these peculiar circumstances so strikingly combined. Either their food is more liable to failure, or they have not sufficient power of wing to search for it over an extensive area, or during some season of the year it becomes very scarce, and less wholesome substitutes have to be found; and thus, though more fertile in offspring, they can never increase beyond the supply of food in the least favourable seasons. Many birds can only exist by migrating, when their food becomes scarce, to regions possessing a milder, or at least a different climate, though, as these migrating birds are seldom excessively abundant, it is evident that the countries they visit are still deficient in a constant and abundant supply of wholesome food. Those whose organization does not permit them to migrate when their food becomes periodically scarce, can never attain a large population. This is probably the reason why woodpeckers are scarce with us, while in the tropics they are among the most abundant of solitary birds. Thus the house sparrow is more abundant than the redbreast, because its food is more constant and plentiful,--seeds of grasses being preserved during the winter, and our farm-yards and stubble-fields furnishing an almost inexhaustible supply. Why, as a general rule, are aquatic, and especially sea birds, very numerous in individuals? Not because they are more prolific than others, generally the contrary; but because their food never fails, the sea-shores and river-banks daily swarming with a fresh supply of small mollusca and crustacea. Exactly the same laws will apply to mammals. Wild cats are prolific and have few enemies; why then are they never as abundant as rabbits? The only intelligible answer is, that their supply of food is more precarious. It appears evident, therefore, that so long as a country remains physically unchanged, the numbers of its animal population cannot materially increase. If one species does so, some others requiring the same kind of food must diminish in proportion. The numbers that die annually must be immense; and as the individual existence of each animal depends upon itself, those that die must be the weakest--the very young, the aged, and the diseased,--while those that prolong their existence can only be the most perfect in health and vigour--those who are best able to obtain food regularly, and avoid their numerous enemies. It is, as we commenced by remarking, "a struggle for existence," in [[p. 57]] which the weakest and least perfectly organized must always succumb.

Now it is clear that what takes place among the individuals of a species must also occur among the several allied species of a group,--viz. that those which are best adapted to obtain a regular supply of food, and to defend themselves against the attacks of their enemies and the vicissitudes of the seasons, must necessarily obtain and preserve a superiority in population; while those species which from some defect of power or organization are the least capable of counteracting the vicissitudes of food, supply, &c., must diminish in numbers, and, in extreme cases, become altogether extinct. Between these extremes the species will present various degrees of capacity for ensuring the means of preserving life; and it is thus we account for the abundance or rarity of species. Our ignorance will generally prevent us from accurately tracing the effects to their causes; but could we become perfectly acquainted with the organization and habits of the various species of animals, and could we measure the capacity of each for performing the different acts necessary to its safety and existence under all the varying circumstances by which it is surrounded, we might be able even to calculate the proportionate abundance of individuals which is the necessary result.

If now we have succeeded in establishing these two points--1st, *that the animal population of a country is generally stationary, being kept down by a periodical deficiency of food, and other checks*; and, 2nd, *that the comparative abundance or scarcity of the individuals of the several species is entirely due to their organization and resulting habits, which, rendering it more difficult to procure a regular supply of food and to provide for their personal safety in some cases than in others, can only be balanced by a difference in the population* [sic] *which have to exist in a given area*--we shall be in a condition to proceed to the consideration of *varieties*, to which the preceding remarks have a direct and very important application.

Most or perhaps all the variations from the typical form of a species must have some definite effect, however slight, on the habits or capacities of the individuals. Even a change of colour might, by rendering them more or less distinguishable, affect their safety; a greater or less development of hair might modify their habits. More important changes, such as an increase in the power or dimensions of the limbs or any of the external organs, would more or less affect their mode of procuring food or the range of [[p. 58]] country which they inhabit. It is also evident that most changes would affect, either favourably or adversely, the powers of prolonging existence. An antelope with shorter or weaker legs must necessarily suffer more from the attacks of the feline carnivora; the passenger pigeon with less powerful wings would sooner or later be affected in its powers of procuring a regular supply of food; and in both cases the result must necessarily be a diminution of the population of the modified species. If, on the other hand, any species should produce a variety having slightly increased powers of preserving existence, that variety must inevitably in time acquire a superiority in numbers. These results must follow as surely as old age, intemperance, or scarcity of food produce an increased mortality. In both cases there may be many individual exceptions; but on the average the rule will invariably be found to hold good. All varieties will therefore fall into two classes--those which under the same conditions would never reach the population of the parent species, and those which would in time obtain and keep a numerical superiority. Now, let some alteration of physical conditions occur in the district--a long period of drought, a destruction of vegetation by locusts, the irruption of some new carnivorous animal seeking "pastures new"--any change in fact tending to render existence more difficult to the species in question, and tasking its utmost powers to avoid complete extermination; it is evident that, of all the individuals composing the species, those forming the least numerous and most feebly organized variety would suffer first, and, were the pressure severe, must soon become extinct. The same causes continuing in action, the parent species would next suffer, would gradually diminish in numbers, and with a recurrence of similar unfavourable conditions might also become extinct. The superior variety would then alone remain, and on a return to favourable circumstances would rapidly increase in numbers and occupy the place of the extinct species and variety.

The *variety* would now have replaced the *species*, of which it would be a more perfectly developed and more highly organized form. It would be in all respects better adapted to secure its safety, and to prolong its individual existence and that of the race. Such a variety *could not* return to the original form; for that form is an inferior one, and could never compete with it for existence. Granted, therefore, a "tendency" to reproduce the original type of the species, still the variety must ever remain preponderant in numbers, and under adverse physical conditions *again alone survive*. [[p. 59]] But this new, improved, and populous race might itself, in course of time, give rise to new varieties, exhibiting several diverging modifications of form, any of which, tending to increase the facilities for preserving existence, must, by the same general law, in their turn become predominant. Here, then, we have *progression and continued divergence* deduced from the general laws which regulate the existence of animals in a state of nature, and from the undisputed fact that varieties do frequently occur. It is not, however, contended that this result would be invariable; a change of physical conditions in the district might at times materially modify it, rendering the race which had been the most capable of supporting existence under the former conditions now the least so, and even causing the extinction of the newer and, for a time, superior race, while the old or parent species and its first inferior varieties continued to flourish. Variations in unimportant parts might also occur, having no perceptible effect on the life-preserving powers; and the varieties so furnished might run a course parallel with the parent species, either giving rise to further variations or returning to the former type. All we argue for is, that certain varieties have a tendency to maintain their existence longer than the original species, and this tendency must make itself felt; for though the doctrine of chances or averages can never be trusted to on a limited scale, yet, if applied to high numbers, the results come nearer to what theory demands, and, as we approach to an infinity of examples, become strictly accurate. Now the scale on which nature works is so vast--the numbers of individuals and periods of time with which she deals approach so near to infinity, that any cause, however slight, and however liable to be veiled and counteracted by accidental circumstances, must in the end produce its full legitimate results.

Let us now turn to domesticated animals, and inquire how varieties produced among them are affected by the principles here enunciated. The essential difference in the condition of wild and domestic animals is this,--that among the former, their well-being and very existence depend upon the full exercise and healthy condition of all their senses and physical powers, whereas, among the latter, these are only partially exercised, and in some cases are absolutely unused. A wild animal has to search, and often to labour, for every mouthful of food--to exercise sight, hearing, and smell in seeking it, and in avoiding dangers, in procuring shelter from the inclemency of the seasons, and in providing for the subsistence and safety of its offspring. There is no muscle of its body that is not called into daily and hourly activity; there is no sense or faculty that is not strengthened by continual exercise. The domestic animal, on the other hand, has food provided for it, is sheltered, and often confined, to guard it against the vicissitudes of the seasons, is carefully secured from the attacks of its natural enemies, and seldom even rears its young without human assistance. Half of its senses and faculties are quite useless; and the other half are but occasionally called into feeble exercise, while even its muscular system is only irregularly called into action.

Now when a variety of such an animal occurs, having increased power or capacity in any organ or sense, such increase is totally useless, is never called into action, and may even exist without the animal ever becoming aware of it. In the wild animal, on the contrary, all its faculties and powers being brought into full action for the necessities of existence, any increase becomes immediately available, is strengthened by exercise, and must even slightly modify the food, the habits, and the whole economy of the race. It creates as it were a new animal, one of superior powers, and which will necessarily increase in numbers and outlive those inferior to it.

Again, in the domesticated animal all variations have an equal chance of continuance; and those which would decidedly render a wild animal unable to compete with its fellows and continue its existence are no disadvantage whatever in a state of domesticity. Our quickly fattening pigs, short-legged sheep, pouter pigeons, and poodle dogs could never have come into existence in a state of nature, because the very first step towards such inferior forms would have led to the rapid extinction of the race; still less could they now exist in competition with their wild allies. The great speed but slight endurance of the race horse, the unwieldy strength of the ploughman's team, would both be useless in a state of nature. If turned wild on the pampas, such animals would probably soon become extinct, or under favourable circumstances might each lose those extreme qualities which would never be called into action, and in a few generations would revert to a common type, which must be that in which the various powers and faculties are so proportioned to each other as to be best adapted to procure food and secure safety,--that in which by the full exercise of every part of his organization the animal can alone continue to live. Domestic varieties, when turned wild, *must* return to something near the type of the original wild stock, *or become altogether extinct.*[3]

We see, then, that no inferences as to varieties in a state of nature can be deduced from the observation of those occurring among domestic animals. The two are so much opposed to each other in every circumstance of their existence, that what applies to the one is almost sure not to apply to the other. Domestic animals are abnormal, irregular, artificial; they are subject to varieties which never occur and never can occur in a state of nature: their very existence depends altogether on human care; so far are many of them removed from that just proportion of faculties, that true balance of organization, by means of which alone an animal left to its own resources can preserve its existence and continue its race.

The hypothesis of Lamarck--that progressive changes in species have been produced by the attempts of animals to increase the development of their own organs, and thus modify their structure and habits--has been repeatedly and easily refuted by all writers on the subject of varieties and species, and it seems to have been considered that when this was done the whole question has been finally settled; but the view here developed renders such an hypothesis quite unnecessary, by showing that similar results must be produced by the action of principles constantly at work in nature. The powerful retractile talons of the falcon- and the cat-tribes have not been produced or increased by the volition of those animals; but among the different varieties which occurred in the earlier and less highly organized forms of these groups, *those always survived longest which had the greatest facilities for seizing their prey.* Neither did the giraffe acquire its long neck by desiring to reach the foliage of the more lofty shrubs, and constantly stretching its neck for the purpose, but because any varieties which occurred among its antitypes with a longer neck than usual *at once secured a fresh range of pasture over the same ground as their shorter-necked companions, and on the first scarcity of food were thereby enabled to outlive them.* Even the peculiar colours of many animals, especially insects, so closely resembling the soil or the leaves or the trunks on which they habitually reside, are explained on the same principle; for though in the course of ages varieties of many tints may have occurred, *yet those races having colours best adapted to concealment from their enemies would inevitably survive the longest.* We have also here an acting cause to account for that balance so often observed in nature,--a deficiency in one set of organs always being compensated by an increased development of some others--powerful wings accompanying weak [[p. 62]] feet, or great velocity making up for the absence of defensive weapons; for it has been shown that all varieties in which an unbalanced deficiency occurred could not long continue their existence. The action of this principle is exactly like that of the centrifugal governor of the steam engine, which checks and corrects any irregularities almost before they become evident; and in like manner no unbalanced deficiency in the animal kingdom can ever reach any conspicuous magnitude, because it would make itself felt at the very first step, by rendering existence difficult and extinction almost sure soon to follow. An origin such as is here advocated will also agree with the peculiar character of the modifications of form and structure which obtain in organized beings--the many lines of divergence from a central type, the increasing efficiency and power of a particular organ through a succession of allied species, and the remarkable persistence of unimportant parts such as colour, texture of plumage and hair, form of horns or crests, through a series of species differing considerably in more essential characters. It also furnishes us with a reason for that "more specialized structure" which Professor Owen states to be a characteristic of recent compared with extinct forms, and which would evidently be the result of the progressive modification of any organ applied to a special purpose in the animal economy.

We believe we have now shown that there is a tendency in nature to the continued progression of certain classes of *varieties* further and further from the original type--a progression to which there appears no reason to assign any definite limits--and that the same principle which produces this result in a state of nature will also explain why domestic varieties have a tendency to revert to the original type. This progression, by minute steps, in various directions, but always checked and balanced by the necessary conditions, subject to which alone existence can be preserved, may, it is believed, be followed out so as to agree with all the phenomena presented by organized beings, their extinction and succession in past ages, and all the extraordinary modifications of form, instinct, and habits which they exhibit.

Ternate, February, 1858.

MSAC Philosophy Group Circular

| Mt. San Antonio College | Walnut, California |

Naturally Selected

Original Papers in Biology

Issue Two

Experiments in Plant Hybridization

Gregor Mendel

a simplified and edited translation

1] Introductory Remarks

In this paper, I will describe experiments in which I artificially fertilized plants. When I got the same results over and over again, I did further experiments. Some people spend their whole lives studying this branch of science. To do this for your whole life requires lots of courage to keep going when it gets tough or boring. To study that long is the only way to discover the secrets of evolution. This paper will now present the results of an experiment. This experiment was done on a certain group of plants, for eight years until I reached a conclusion. The reader must decide whether the results are correct or not.

[2] Selection of the Experimental Plants

To have a successful experiment, you need to choose the right plants most suitable for the experiment. The selection of the plants must be done very carefully so as to not risk anything or get incorrect results. The plants must:

1. Have different traits.

2. The hybrids of the plants must, during the flowering period, be protected from bugs that could pollinate them and create incorrect results.

3. The hybrids and their offspring should always be able to produce seeds that can be grown into more plants.

If bugs did pollinate the plants, it would have a bad effect on the results. If the plants could no longer produce seeds that could be grown into more plants, the experiments would be very difficult and it could even end them. To determine the relationship between plants, the plants must be closely watched at all times. 34 different pea plants were gotten and were put into a two-year trial. They were allowed to fertilize themselves and 22 were selected that always produced offspring that looked exactly like the parent plant.

[3] Division and Arrangement of the Experiments

Suppose you have two plants that are different for one particular characteristic. The goal of this experiment was to find out why, when the two plants are crossed, only one characteristic shows up. The characteristics that were selected for these experiments were:

1. The shape of the ripe seeds: either round or wrinkled.

2. The color of the inside of the seed: either pale yellow, bright yellow or orange colored.

3. The color of the seed coat: either white or grayish brown.

4. The shape of the ripe pods: either puffed up or wrinkled and squished.

5. The color of the unripe pods: either green or yellow. *

6. The position of the flowers on the stem: either distributed along the main stem or bunched at the top of the stem.

7. The length of the stem: either tall or short.

A plant with each characteristic was crossed with a plant with the opposite characteristic, and the following numbers of fertilizations were done:

1st experiment plants.	60 fertilizations on 15
2nd experiment plants.	58 fertilizations on 10
3rd experiment plants.	35 fertilizations on 10
4th experiment plants.	40 fertilizations on 10
5th experiment plants.	23 fertilizations on 5
6th experiment plants.	34 fertilizations on 10
7th experiment plants.	37 fertilizations on 10

* One pea plant group had a brownish-red colored pod, which turned violet or blue when it ripened.

The plants were grown in garden beds, and a few in pots, and were kept upright using sticks, branches of trees, and strings stretched between sticks. For each experiment a number of potted plants were put in a greenhouse, to make sure the results were correct and to make sure that bugs didn't pollinate them.

[4] The Forms of the Hybrids

In other experiments, it was found that if you crossed, for example, a tall plant and a short plant, the hybrids produced were not medium-sized, they were tall! Peas are exactly the same. Only one characteristic is shown. The "shown" characteristic is called dominating and the "unshown" haracteristic is called recessive. The characteristics that were dominant in these experiments were:

1. The round shape of the seed.

2. The yellow color of the inside of the seed.

3. The grayish-brown color of the seed coat.

4. The puffed up shape of the pod.

5. The green color of the unripe pod.

6. The distribution of the flowers along the main stem.

7. The tall stem.

[5] The First Generation from the Hybrids

In this experiment, the hybrids produced in earlier experiments were allowed to self-fertilize. In this generation the dominant traits reappeared, along with the recessive ones. If there were four offspring plants, three would show a dominant characteristic and one would show a recessive characteristic, making a 3:1 RATIO.

The characteristics that were recessive were:

1. 1.   The wrinkled seed.
2. 2.   The green color of the inside of the seed.
3. 3.   The white color of the seed coat.
4. 4.   The wrinkled, squished shape of the pod.
5. 5.   The yellow color of the unripe pod.
6. 6.   The flowers bunched at the top of the stem.
7. 7.   The short stem.

* Exp. 1: Shape of the seed: 253 plants and 7324 seeds were gotten. There were 5474 round seeds and 1850 wrinkled seeds. The ratio was 2.96:1.

* Exp. 2: Color of the inside of the seed: 258 plants and 8023 seeds = 6022 yellow and 2001 green. The ratio was 3.01:1.

	Experiment 1		Experiment 2	
	Shape of Seed		Color of Inside of Seed	
Plants	Round	Angular	Yellow	Green
1	45	12	25	11
2	27	8	32	7
3	24	7	14	5
4	19	10	70	27
5	32	11	24	13
7	88	24	32	13
8	22	10	44	9
9	28	6	50	14
10	25	7	44	18

* Exp. 3: Color of the seed coat: Among 929 plants, 705 had grayish-brown seed coats and 224 had white seed coats, giving a ratio of 3.15:1.

* Exp. 4: Shape of the pods: Of 1181 plants, 882 of them had puffed up pods and 299 had wrinkled ones. The resulting ratio was 2.95:1.

* Exp. 5: Color of the unripe pods: 428 had green pods and 152 had yellow ones, giving a ratio of 2.82:1.

* Exp. 6: Position of flowers on the stem: Among 858 plants, 651 had flowers distributed evenly on the stem and 207 had flowers only at the top, giving a ratio of 3.14:1.

* Exp. 7: Length of stem: Out of 1064 plants, 787 had a tall stem, and 277 had a short stem, giving a ratio of 2.84:1.

If the results of the whole experiment were brought together, the average ratio was 2.98:1, or 3:1.

[6] The Second Generation from the Hybrids

The plants from the previous experiment (the offspring produced when the hybrids self-fertilized) were allowed to self-fertilize. The plants that showed the recessive characteristic produced offspring that were also recessive. Of those plants that had the dominant character, two thirds of them actually were hybrids and had both the dominant and recessive character, and one third had only the dominant character, for a ratio of 2:1. Those plants that had only the dominant character produced offspring that all showed the dominant characteristic. The remaining hybrid plants produced some offspring that showed the dominant characteristic and some that showed the recessive characteristic, in a ratio of 3 dominant to one recessive, just like when their hybrid parents were allowed to self-fertilize. The separate experiments in which the offspring that showed the dominant characteristic were allowed to self-fertilize had these results:

* Exp. 1: Among 565 plants that were raised from round seeds, 193 produced round seeds only, and 372 produced some plants with round seeds and some plants with wrinkled seeds, in a proportion of 3:1.

* Exp. 2: Of 519 plants that were raised from seeds for which the inside of the seed was yellow, 166 produced seeds that were yellow inside and 353 produced some seeds whose insides were yellow and some seeds whose insides were green, in a ratio of 3:1.

* Exp. 3: The offspring of 36 plants produced only gray-brown seed coats, while some of the offspring of 64 plants produced grayish-brown seed coats and some produced white seed coats..

* Exp. 4: The offspring of 29 plants produced only puffed up pods while some of the offspring of 71 plants had puffed up pods and some had wrinkled pods.

* Exp. 5: The offspring of 40 plants produced only green pods, while some of the offspring of 60 plants produced green pods and some produced yellow ones.

* Exp. 6.: The offspring of 33 plants had only flowers distributed on the stem, while some of the offspring of 67 had flowers distributed on the stem and some had flowers bunched on the top.

* Exp. 7: The offspring of 28 plants had only tall stems, while some of the offspring of 72 plants had tall stems and some had short stems.

[7] The Subsequent Generations from the Hybrids

Summary: The offspring produced when hybrids self-fertilized can be classified as either hybrids (possessing both the dominant and the recessive character), dominant (possessing only the dominant character), or recessive (possessing only the recessive character). The ratio was 2 hybrids to one dominant to one recessive (2:1:1 ratio). If **A** refers to one of the two constant characters, for instance the dominant character, and **a** refers to the recessive characters, and **Aa** refers to the hybrid form in which both characters are present, the expression

$$A + 2Aa + a$$

shows the proportion of offspring of the self-fertilizing hybrids.

[8] The Offspring of Hybrids in which Several Differentiating Characters are Associated.

Summary: Experiments were done in which plants were crossed that differed in two or three characters. For example, crosses were done among plants that differed in all three of the following characters: seed coat form (round or wrinkled), color of the inside of the seed (green or yellow) and seed coat color (grayish-brown or white).

[9] The Reproductive Cells of the Hybrids

Summary: Each parent plant had two factors that were able to pass on traits from parent to offspring, but that only one of those factors was present in the egg cell or the pollen cell. So hybrid parent plants produced eggs or pollen, and equal numbers of the egg or pollen cells had the dominant character and the recessive character.

Separate experiments proved that this was true of both pollen and egg cells. It is purely a matter of chance, which of the two sorts of pollen will become united with each separate egg cell.

So when crossing two hybrids (Aa X Aa), the hybrids are able to produce, besides the two parental forms, offspring that are like themselves. A/a and a/A both give the same union Aa, since it makes no difference in the result of fertilization to which of the two characters the pollen or egg cells belong. So A/A + A/a + a/A + a/a = A + 2 Aa + a.

[10] Experiments with Hybrids of Other Species of Plants

Summary: Other experiments were used to determine whether the laws discovered for peas applied to the hybrids of other plants. Several experiments were done with different species of beans. With some crosses of beans, looking at traits that determine the form of the plants, the ratios of the numbers in which the different forms appeared in the separate generations were the same as with peas. But in other crosses with other species of beans, in which flower color was examined, the results were confusing.

[11] Concluding Remarks

Summary: With peas it was shown by experiment that the hybrids form egg and pollen cells of different kinds, and that this holds the key to the variability of their offspring. So in peas, it is beyond a doubt that when a new embryo is formed, a perfect union of the elements of both reproductive cells must take place. How could we otherwise explain that among the offspring of the hybrids, both original types reappear in equal numbers and with all their peculiarities?

But even the truth of the law formulated for peas must still be confirmed and some of the more important experiments must be repeated. It is important to compare the observations made regarding peas with the results arrived at by the two authorities in this branch of knowledge, Köreuter and Gärtner, in their investigations. A description follows of experiments carried out by Kölreuter, Gärtner, and others, including experiments made with respect to the transformation of one species into another by artificial fertilization.

Excerpted from:

http://web.pdx.edu/~cruzan/Kid's%20Mendel%20Web/Menu.htm

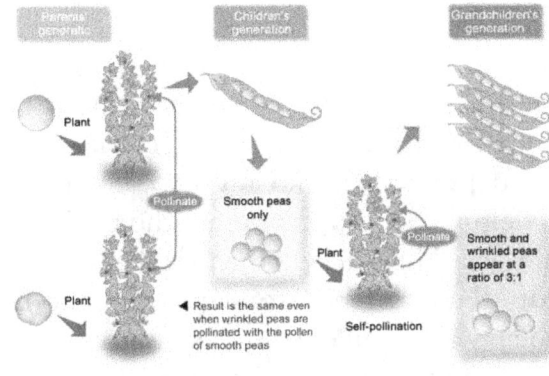

GREGOR MENDEL

Researcher: **Rachel Sahlman**

Johann Mendel, who changed his name to Gregor when he entered a monastery as a young man, was born on his father's farm on July 22, 1822. By the time he was 11 years old, Johann had learned everything possible from his local school. The schoolmaster suggested that he attend a school in a larger town many miles away. This would be difficult for Mendel's family, because he would need to pay for room and board.

Mendel's father owned a 45-acre farm; however, the family lived a frugal life with little money for "extras." Nevertheless, Mendel's family found a way to send him to Leipnik School, where he studied for one year. His teachers then recommended that Mendel attend Troppau High School, which was 20 miles from his home. Because money was tight, Mendel often went hungry. However, he finished his six-year course in 1840 and graduated with high honors.

Mendel realized that he needed additional schooling to begin a career. Upon graduation, he enrolled at the Philosophical Institute at Olmütz, which was 100 miles away. After two years at the Institute, Mendel was unsure of his future. His physics professor recommended that Mendel join St. Thomas Monastery in Brünn, Austria, now known as Brno, Czech Republic. At the monastery, Mendel did not have to worry about food or shelter, and he received a fine education. In 1843, Mendel was accepted as a trial member at the monastery, and he took his new name, Gregor.

In 1848, Mendel was assigned as a priest, but he soon realized that parish work was not for him. He was reassigned to a high school teaching position and was considered a good teacher. However, Mendel was not able to pass the teaching exam, and it was recommended that he spend two years at the University of Vienna at the monastery's expense. Mendel completed his studies, but again he was unable to pass the exam.

Mendel continued to teach on a temporary basis, but he spent the majority of his time in the monastery gardens. He developed an interest in what we now call heredity, and he experimented with pea plants in an attempt to prove his theories. Mendel decided to cross-fertilize plants with opposite characteristics; such as tall plants with short plants, and plants that produced smooth peas with those that produced wrinkled peas. Through these experiments, Mendel was able to set out a theory that all living things have aspects called dominant and recessive traits.

In February 1865, Mendel presented his findings to the Natural History Society of Brünn and published them under the title "Experiments in Plant Hybridization" in a scientific journal. No one seemed interested in Mendel's findings. Even when he sent his papers to renowned University of Munich professor Karl von Nägeli, his discoveries went unappreciated. The years passed, and Mendel realized that his findings were not going to be recognized in his lifetime.

In 1868, Mendel was elected Abbot and Prelate of St.Thomas Monastery, which left him little time to continue his studies and experiments. He led a comfortable life for the next 15 years. Gregor Mendel died on January 6, 1884 at the age of 62. Sixteen years later, Mendel's findings were rediscovered.

In 1900, three botanists in different parts of Europe came across his papers. Hugo de Vries, Carl Correns, and Erich Tschermak praised Mendel's research and achieved similar results in their own studies. Mendel's principles of heredity began to be referred to as the Mendelian Laws, and these laws are considered to be the foundation of the modern study of genetics.

MENDEL'S CONTRIBUTION TO THE HISTORY OF GENETICS

Genetics as a discipline is young, but the concept that forms its subject—inheritance—stretches back in time. The word has been formed from the adjective *genetic,* found in the sciences of the nineteenth and twentieth centuries—for example, biogenetic law, genetic affinity, genetic psychology—and meaning, according to the *Oxford English Dictionary,* "pertaining to, or having references to, origin." Not until 1906 was the noun *genetics* publicly proposed to cover those labors that, in the words of its author, William Bateson, "are devoted to the elucidation of the phenomena of heredity and variation: in other words to the physiology of descent, with implied bearing on the theoretical problems of the evolutionist and the systematist, and application to the practical problems of breeders, whether of animals or plants."

We begin, therefore, by commenting briefly on the long history of notions of heredity and variation, reflecting the while on the significance of cultural and economic factors that both drew attention to them and shaped them. Next we turn to the father figure of genetics, Gregor Mendel (1822–1884), and his introduction of the Mendelian experiment. With the rediscovery of his work in 1900 we explore the contribution of the early "Mendelians," the melding of Mendelian heredity with the theory of the chromosome, a synthesis that revealed a geography of heredity in the cell but did not answer the question, "What is the identity of the genetic material?" Only with the introduction of the Watson-Crick structure for DNA in 1953 was a window opened through which to glimpse the terrain that was to become "molecular genetics."

Heredity and variation are today considered as two sides of the same coin. Thus variation among sibs results from the varied commingling and expression of the hereditary determinants from two parents. Spontaneous changes in the hereditary material (mutations) give rise to variations, and these are inherited. In other words there is heredity, variation, and the heredity of variation, and they belong together. Prior to the latter half of the nineteenth century this conceptual framework did not exist. There still lingered relics of the ancient view of heredity as the result of the reproduction of the type, whereas any deviance from the type was ascribed to the effects of the mother's imagination, changed conditions of life, and so on. Moreover both heredity and variation were situated within the broad topic of "generation." This included the regeneration of lost and damaged organs, the development of the embryo from the fertilized egg, and all forms of reproduction, both sexual and asexual. But as attention directed to human heredity increased during the nineteenth century the subject of the transmission of hereditary characters began to acquire a separate conceptual status.

In the eighteenth century two rival conceptions of the phenomenon of inheritance were in play: the doctrine of preformation claimed that like generates like because the offspring are in some way already present in the germ and have only to unfold or "evolve" to yield offspring like the parents. A literal view of this doctrine was held by some who claimed that hidden miniatures of future generations must have existed from the time of creation, all nested one within another, like a set of Russian dolls, hence the term preexistence for this version of the doctrine. Procreation became, as it were, the act of revealing what had been created long before. Opposed to this was the doctrine of epigenesis, according to which the parts of the offspring are formed successively, and in the process the embryo undergoes a series of transformations. Support for this view came from descriptions of the development of the fertilized egg. Experimental support came also from the hybridization experiments of Joseph Kölreuter, who in 1766 described how he had succeeded in transforming one species of plant into another. How, he asked, was this possible if the offspring is already preformed? He also found that no matter which parent supplied the "male seed" (pollen) and which the female "seed" (ovules), the resulting hybrids were the same. How could these results be accounted for on the basis of any theory of preformation—whether, as the animalculists claimed, the preformed miniatures reside in the spermatozoa, or as the ovists urged, in the egg? A further difficulty arose for preformationists when nineteenth-century microscopists developed a cell theory according to which the fundamental unit of life is the cell, not some preformed rudiments of the organism.

These two resources—hybridization and the cell theory—were to form the basis for Gregor Mendel's research. Prior to his work, however, the study of hybridization had been carried out under different assumptions and in a much more limited manner. The number of hybrid offspring grown in an experiment was limited, and the prevalence of the effect of hybridization in diluting character differences supported the view that heredity is a blending process, like the mixing of ink and water. The conception of the organism too was generally holistic: that is, the type of the species acts as a whole, its "essential" characteristics—flower color, habit, leaf shape, and so on—being but expressions of the type. With this point of view, effective analysis of hereditary transmission is impossible.

Excerpted from:

http://science.jrank.org/pages/7723/History-Genetics.html

MSAC Philosophy Group Circular

| Mt. San Antonio College | Walnut, California |

Naturally Selected

Original Papers in Biology

Issue Three

On the Origin of Species

Thomas H. Huxley

Westminster Review, April 1860

MR. DARWIN'S long-standing and well-earned scientific eminence probably renders him indifferent to that social notoriety which passes by the name of success; but if the calm spirit of the philosopher have not yet wholly superseded the ambition and the vanity of the carnal man within him, he must be well satisfied with the results of his venture in publishing the 'Origin of Species'. Overflowing the narrow bounds of purely scientific circles, the "species question" divides with Italy and the Volunteers the attention of general society. Everybody has read Mr. Darwin's book, or, at least, has given an opinion upon its merits or demerits; pietists, whether lay or ecclesiastic, decry it with the mild railing which sounds so charitable; bigots denounce it with ignorant invective; old ladies of both sexes consider it a decidedly dangerous book, and even savants, who have no better mud to throw, quote antiquated writers to show that its author is no better than an ape himself; while every philosophical thinker hails it as a veritable Whitworth gun in the armoury of liberalism; and all competent naturalists and physiologists, whatever their opinions as to the ultimate fate of the doctrines put forth, acknowledge that the work in which they are embodied is a solid contribution to knowledge and inaugurates a new epoch in natural history.

Nor has the discussion of the subject been restrained within the limits of conversation. When the public is eager and interested, reviewers must minister to its wants; and the genuine 'litterateur' is too much in the habit of acquiring his knowledge from the book he judges—as the Abyssinian is said to provide himself with steaks from the ox which carries him—to be withheld from criticism of a profound scientific work by the mere want of the requisite preliminary scientific acquirement; while, on the other hand, the men of science who wish well to the new views, no less than those who dispute their validity, have naturally sought opportunities of expressing their opinions. Hence it is not surprising that almost all the critical journals have noticed Mr. Darwin's work at greater or less length; and so many disquisitions, of every degree of excellence, from the poor product of ignorance, too often stimulated by prejudice, to the fair and thoughtful essay of the candid student of Nature, have appeared, that it seems an almost hopeless task to attempt to say anything new upon the question.

But it may be doubted if the knowledge and acumen of prejudged scientific opponents, or the subtlety of orthodox special pleaders, have yet exerted their full force in mystifying the real issues of the great controversy which has been set afoot, and whose end is hardly likely to be seen by this generation; so that, at this eleventh hour, and even failing anything new, it may be useful to state afresh that which is true, and to put the fundamental positions advocated by Mr. Darwin in such a form that they may be grasped by those whose special studies lie in other directions. And the adoption of this course may be the more advisable, because, notwithstanding its great deserts, and indeed partly on account of them, the 'Origin of Species' is by no means an easy book to read—if by reading is implied the full comprehension of an author's meaning.

We do not speak jestingly in saying that it is Mr. Darwin's misfortune to know more about the question he has taken up than any man living. Personally and practically exercised in zoology, in minute anatomy, in geology; a student of geographical distribution, not on maps and in museums only, but by long voyages and laborious collection; having largely advanced each of these branches of science, and having spent many years in gathering and sifting materials for his present work, the store of accurately registered facts upon which the author of the 'Origin of Species' is able to draw at will is prodigious.

But this very superabundance of matter must have been embarrassing to a writer who, for the present, can only put forward an abstract of his views; and thence it arises, perhaps, that notwithstanding the clearness of the style, those who attempt fairly to digest the book find much of it a sort of intellectual pemmican—a mass of facts crushed and pounded into shape, rather than held together by the ordinary medium of an obvious logical bond; due attention will, without doubt, discover this bond, but it is often hard to find.

Again, from sheer want of room, much has to be taken for granted which might readily enough be proved; and hence, while the adept, who can supply the missing links in the evidence from his own knowledge, discovers fresh proof of the singular thoroughness with which all difficulties have been considered and all unjustifiable suppositions avoided, at every reperusal of Mr. Darwin's pregnant paragraphs, the novice in biology is apt to complain of the frequency of what he fancies is gratuitous assumption.

Thus while it may be doubted if, for some years, any one is likely to be competent to pronounce judgment on all the issues raised by Mr. Darwin, there is assuredly abundant room for him, who, assuming the humbler, though perhaps as useful, office of an interpreter between the 'Origin of Species' and the public, contents himself with endeavouring to point out the nature of the problems which it discusses; to distinguish between the ascertained facts and the theoretical views which it contains; and finally, to show the extent to which the explanation it offers satisfies the requirements of scientific logic. At any rate, it is this office which we purpose to undertake in the following pages.

It may be safely assumed that our readers have a general conception of the nature of the objects to which the word "species" is applied; but it has, perhaps, occurred to a few, even to those who are naturalists 'ex professo', to reflect, that, as commonly employed, the term has a double sense and denotes two very different orders of relations. When we call a group of animals, or of plants, a species, we may imply thereby, either that all these animals or plants have some common peculiarity of form or structure; or, we may mean that they possess some common functional character. That part of biological science which deals with form and structure is called Morphology—that which concerns itself with function, Physiology—so that we may conveniently speak of these two senses, or aspects, of "species"—the one as morphological, the other as physiological. Regarded from the former point of view, a species is nothing more than a kind of animal or plant, which is distinctly definable from all others, by certain constant, and not merely sexual, morphological peculiarities. Thus horses form a species, because the group of animals to which that name is applied is distinguished from all others in the world by the following constantly associated characters. They have—1, A vertebral column; 2, Mammae; 3, A placental embryo; 4, Four legs; 5, A single well-developed toe in each foot provided with a hoof; 6, A bushy tail; and 7, Callosities on the inner sides of both the fore and the hind legs. The asses, again, form a distinct species, because, with the same characters, as far as the fifth in the above list, all asses have tufted tails, and have callosities only on the inner side of the fore-legs. If animals were discovered having the general characters of the horse, but sometimes with callosities only on the fore-legs, and more or less tufted tails; or animals having the general characters of the ass, but with more or less bushy tails, and sometimes with callosities on both pairs of legs, besides being intermediate in other respects—the two species would have to be merged into one. They could no longer be regarded as morphologically distinct species, for they would not be distinctly definable one from the other.

However bare and simple this definition of species may appear to be, we confidently appeal to all practical naturalists, whether zoologists, botanists, or palaeontologists, to say if, in the vast majority of cases, they know, or mean to affirm anything more of the group of animals or plants they so denominate than what has just been stated. Even the most decided advocates of the received doctrines respecting species admit this.

"I apprehend," says Professor Owen, "that few naturalists nowadays, in describing and proposing a name for what they call 'a new species,' use that term to signify what was meant by it twenty or thirty years ago; that is, an originally distinct creation, maintaining its primitive distinction by obstructive generative peculiarities. The proposer of the new species now intends to state no more than he actually knows; as, for example, that the differences on which he founds the specific character are constant in individuals of both sexes, so far as observation has reached; and that they are not due to domestication or to artificially superinduced external circumstances, or to any outward influence within his cognizance; that the species is wild, or is such as it appears by Nature."

If we consider, in fact, that by far the largest proportion of recorded existing species are known only by the study of their skins, or bones, or other lifeless exuvia; that we are acquainted with none, or next to none, of their physiological peculiarities, beyond those which can be deduced from their structure, or are open to cursory observation; and that we cannot hope to learn more of any of those extinct forms of life which now constitute no inconsiderable proportion of the known Flora and Fauna of the world: it is obvious that the definitions of these species can be only of a purely structural, or morphological, character. It is probable that naturalists would have avoided much confusion of ideas if they had more frequently borne the necessary limitations of our knowledge in mind. But while it may safely be admitted that we are acquainted with only the morphological characters of the vast majority of species—the functional or physiological, peculiarities of a few have been carefully investigated, and the result of that study forms a large and most interesting portion of the physiology of reproduction.

The student of Nature wonders the more and is astonished the less, the more conversant he becomes with her operations; but of all the perennial miracles she offers to his inspection, perhaps the most worthy of admiration is the development of a plant or of an animal from its embryo. Examine the recently laid egg of some common animal, such as a salamander or newt. It is a minute spheroid in which the best microscope will reveal nothing but a structureless sac, enclosing a glairy fluid, holding granules in suspension. But strange possibilities lie dormant in that semi-fluid globule. Let a moderate supply of warmth reach its watery cradle, and the plastic matter undergoes changes so rapid, yet so steady and purposelike in their succession, that one can only compare them to those operated by a skilled modeller upon a formless lump of clay. As with an invisible trowel, the mass is divided and subdivided into smaller and smaller portions, until it is reduced to an aggregation of granules not too large to build withal the finest fabrics of the nascent organism. And, then, it is as if a delicate finger traced out the line to be occupied by the spinal column, and moulded the contour of the body; pinching up the head at one end, the tail at the other, and fashioning flank and limb into due salamandrine proportions, in so artistic a way, that, after watching the process hour by hour, one is almost involuntarily possessed by the notion, that some more subtle aid to vision than an achromatic, would show the hidden artist, with his plan before him, striving with skilful manipulation to perfect his work.

As life advances, and the young amphibian ranges the waters, the terror of his insect contemporaries, not only are the nutritious particles supplied by its prey, by the addition of which to its frame, growth takes place, laid down, each in its proper spot, and in such due proportion to the rest, as to reproduce the form, the colour, and the size, characteristic of the parental stock; but even the wonderful powers of reproducing lost parts possessed by these animals are controlled by the same governing tendency. Cut off the legs, the tail, the jaws, separately or all together, and, as Spallanzani showed long ago, these parts not only grow again, but the redintegrated limb is formed on the same type as those which were lost. The new jaw, or leg, is a newt's, and never by any accident more like that of a frog. What is true of the newt is true of every animal and of every plant; the acorn tends to build itself up again into a woodland giant such as that from whose twig it fell; the spore of the humblest lichen reproduces the green or brown incrustation which gave it birth; and at the other end of the scale of life, the child that resembled neither the paternal nor the maternal side of the house would be regarded as a kind of monster.

So that the one end to which, in all living beings, the formative impulse is tending—the one scheme which the Archaeus of the old speculators strives to carry out, seems to be to mould the offspring into the likeness of the parent. It is the first great law of reproduction, that the offspring tends to resemble its parent or parents, more closely than anything else.

Science will some day show us how this law is a necessary consequence of the more general laws which govern matter; but, for the present, more can hardly be said than that it appears to be in harmony with them. We know that the phenomena of vitality are not something apart from other physical phenomena, but one with them; and matter and force are the two names of the one artist who fashions the living as well as the lifeless. Hence living bodies should obey the same great laws as other matter—nor, throughout Nature, is there a law of wider application than this, that a body impelled by two forces takes the direction of their resultant. But living bodies may be regarded as nothing but extremely complex bundles of forces held in a mass of matter, as the complex forces of a magnet are held in the steel by its coercive force; and, since the differences of sex are comparatively slight, or, in other words, the sum of the forces in each has a very similar tendency, their resultant, the offspring, may reasonably be expected to deviate but little from a course parallel to either, or to both.

Represent the reason of the law to ourselves by what physical metaphor or analogy we will, however, the great matter is to apprehend its existence and the importance of the consequences deducible from it. For things which are like to the same are like to one another; and if, in a great series of generations, every offspring is like its parent, it follows that all the offspring and all the parents must be like one another; and that, given an original parental stock, with the opportunity of undisturbed multiplication, the law in question necessitates the production, in course of time, of an indefinitely large group, the whole of whose members are at once very similar and are blood relations, having descended from the same parent, or pair of parents. The proof that all the members of any given group of animals, or plants, had thus descended, would be ordinarily considered sufficient to entitle them to the rank of physiological species, for most physiologists consider species to be definable as "the offspring of a single primitive stock."

But though it is quite true that all those groups we call species 'may', according to the known laws of reproduction, have descended from a single stock, and though it is very likely they really have done so, yet this conclusion rests on deduction and can hardly hope to establish itself upon a basis of observation. And the primitiveness of the supposed single stock, which, after all, is the essential part of the matter, is not only a hypothesis, but one which has not a shadow of foundation, if by "primitive" he meant "independent of any other living being." A scientific definition, of which an unwarrantable hypothesis forms an essential part, carries its condemnation within itself; but, even supposing such a definition were, in form, tenable, the physiologist who should attempt to apply it in Nature would soon find himself involved in great, if not inextricable, difficulties. As we have said, it is indubitable that offspring 'tend' to resemble the parental organism, but it is equally true that the similarity attained never amounts to identity, either in form or in structure. There is always a certain amount of deviation, not only from the precise characters of a single parent, but when, as in most animals and many plants, the sexes are lodged in distinct individuals, from an exact mean between the two parents. And indeed, on general principles, this slight deviation seems as intelligible as the general similarity, if we reflect how complex the co-operating "bundles of forces" are, and how improbable it is that, in any case, their true resultant shall coincide with any mean between the more obvious characters of the two parents. Whatever be its cause, however, the co-existence of this tendency to minor variation with the tendency to general similarity, is of vast importance in its bearing on the question of the origin of species.

As a general rule, the extent to which an offspring differs from its parent is slight enough; but, occasionally, the amount of difference is much more strongly marked, and then the divergent offspring receives the name of a Variety. Multitudes, of what there is every reason to believe are such varieties, are known, but the origin of very few has been accurately recorded, and of these we will select two as more especially illustrative of the main features of variation. The first of them is that of the "Ancon," or "Otter" sheep, of which a careful account is given by Colonel David Humphreys, F.R.S., in a letter to Sir Joseph Banks, published in the Philosophical Transactions for 1813. It appears that one Seth Wright, the proprietor of a farm on the banks of the Charles River, in Massachusetts, possessed a flock of fifteen ewes and a ram of the ordinary kind. In the year 1791, one of the ewes presented her owner with a male lamb, differing, for no assignable reason, from its parents by a proportionally long body and short bandy legs, whence it was unable to emulate its relatives in those sportive leaps over the neighbours' fences, in which they were in the habit of indulging, much to the good farmer's vexation.

The second case is that detailed by a no less unexceptionable authority than Reaumur, in his 'Art de faire eclore les Poulets'. A Maltese couple, named Kelleia, whose hands and feet were constructed upon the ordinary human model, had born to them a son, Gratio, who possessed six perfectly movable fingers on each hand, and six toes, not quite so well formed, on each foot. No cause could be assigned for the appearance of this unusual variety of the human species.

Two circumstances are well worthy of remark in both these cases. In each, the variety appears to have arisen in full force, and, as it were, 'per saltum'; a wide and definite difference appearing, at once, between the Ancon ram and the ordinary sheep; between the six-fingered and six-toed Gratio Kelleia and ordinary men. In neither case is it possible to point out any obvious reason for the appearance of the variety. Doubtless there were determining causes for these as for all other phenomena; but they do not appear, and we can be tolerably certain that what are ordinarily understood as changes in physical conditions, as in climate, in food, or the like, did not take place and had nothing to do with the matter. It was no case of what is commonly called adaptation to circumstances; but, to use a conveniently erroneous phrase, the variations arose spontaneously. The fruitless search after final causes leads their pursuers a long way; but even those hardy teleologists, who are ready to break through all the laws of physics in chase of their favourite will-o'-the-wisp, may be puzzled to discover what purpose could be attained by the stunted legs of Seth Wright's ram or the hexadactyle members of Gratio Kelleia.

Varieties then arise we know not why; and it is more than probable that the majority of varieties have arisen in this "spontaneous" manner, though we are, of course, far from denying that they may be traced, in some cases, to distinct external influences; which are assuredly competent to alter the character of the tegumentary covering, to change colour, to increase or diminish the size of muscles, to modify constitution, and, among plants, to give rise to the metamorphosis of stamens into petals, and so forth. But however they may have arisen, what especially interests us at present is, to remark that, once in existence, varieties obey the fundamental law of reproduction that like tends to produce like; and their offspring exemplify it by tending to exhibit the same deviation from the parental stock as themselves. Indeed, there seems to be, in many instances, a pre-potent influence about a newly-arisen variety which gives it what one may call an unfair advantage over the normal descendants from the same stock. This is strikingly exemplified by the case of Gratio Kelleia, who married a woman with the ordinary pentadactyle extremities, and had by her four children, Salvator, George, Andre, and Marie. Of these children Salvator, the eldest boy, had six fingers and six toes, like his father; the second and third, also boys, had five fingers and five toes, like their mother, though the hands and feet of George were slightly deformed. The last, a girl, had five fingers and five toes, but the thumbs were slightly deformed. The variety thus reproduced itself purely in the eldest, while the normal type reproduced itself purely in the third, and almost purely in the second and last: so that it would seem, at first, as if the normal type were more powerful than the variety. But all these children grew up and intermarried with normal wives and husband, and then, note what took place: Salvator had four children, three of whom exhibited the hexadactyle members of their grandfather and father, while the youngest had the pentadactyle limbs of the mother and grandmother; so that here, notwithstanding a double pentadactyle dilution of the blood, the hexadactyle variety had the best of it. The same pre-potency of the variety was still more markedly exemplified in the progeny of two of the other children, Marie and

George. Marie (whose thumbs only were deformed) gave birth to a boy with six toes, and three other normally formed children; but George, who was not quite so pure a pentadactyle, begot, first, two girls, each of whom had six fingers and toes; then a girl with six fingers on each hand and six toes on the right foot, but only five toes on the left; and lastly, a boy with only five fingers and toes. In these instances, therefore, the variety, as it were, leaped over one generation to reproduce itself in full force in the next. Finally, the purely pentadactyle Andre was the father of many children, not one of whom departed from the normal parental type.

If a variation which approaches the nature of a monstrosity can strive thus forcibly to reproduce itself, it is not wonderful that less aberrant modifications should tend to be preserved even more strongly; and the history of the Ancon sheep is, in this respect, particularly instructive. With the "'cuteness" characteristic of their nation, the neighbours of the Massachusetts farmer imagined it would be an excellent thing if all his sheep were imbued with the stay-at-home tendencies enforced by Nature upon the newly-arrived ram; and they advised Wright to kill the old patriarch of his fold, and install the Ancon ram in his place. The result justified their sagacious anticipations, and coincided very nearly with what occurred to the progeny of Gratio Kelleia. The young lambs were almost always either pure Ancons, or pure ordinary sheep. [3] But when sufficient Ancon sheep were obtained to interbreed with one another, it was found that the offspring was always pure Ancon. Colonel Humphreys, in fact, states that he was acquainted with only "one questionable case of a contrary nature." Here, then, is a remarkable and well-established instance, not only of a very distinct race being established 'per saltum', but of that race breeding "true" at once, and showing no mixed forms, even when crossed with another breed.

By taking care to select Ancons of both sexes, for breeding from, it thus became easy to establish an extremely well-marked race; so peculiar that, even when herded with other sheep, it was noted that the Ancons kept together. And there is every reason to believe that the existence of this breed might have been indefinitely protracted; but the introduction of the Merino sheep, which were not only very superior to the Ancons in wool and meat, but quite as quiet and orderly, led to the complete neglect of the new breed, so that, in 1813, Colonel Humphreys found it difficult to obtain the specimen, whose skeleton was presented to Sir Joseph Banks. We believe that, for many years, no remnant of it has existed in the United States.

Gratio Kelleia was not the progenitor of a race of six-fingered men, as Seth Wright's ram became a nation of Ancon sheep, though the tendency of the variety to perpetuate itself appears to have been fully as strong in the one case as in the other. And the reason of the difference is not far to seek. Seth Wright took care not to weaken the Ancon blood by matching his Ancon ewes with any but males of the same variety, while Gratio Kelleia's sons were too far removed from the patriarchal times to intermarry with their sisters; and his grandchildren seem not to have been attracted by their six-fingered cousins. In other words, in the one example a race was produced, because, for several generations, care was taken to 'select' both parents of the breeding stock from animals exhibiting a tendency to vary in the same condition; while, in the other, no race was evolved, because no such selection was exercised. A race is a propagated variety; and as, by the laws of reproduction, offspring tend to assume the parental forms, they will be more likely to propagate a variation exhibited by both parents than that possessed by only one.

There is no organ of the body of an animal which may not, and does not, occasionally, vary more or less from the normal type; and there is no variation which may not be transmitted and which, if selectively transmitted, may not become the foundation of a race. This great truth, sometimes forgotten by philosophers, has long been familiar to practical agriculturists and breeders; and upon it rest all the methods of improving the breeds of domestic animals, which, for the last century, have been followed with so much success in England. Colour, form, size, texture of hair or wool, proportions of various parts, strength or weakness of constitution, tendency to fatten or to remain lean, to give much or little milk, speed, strength, temper, intelligence, special instincts; there is not one of these characters whose transmission is not an every-day occurrence within the experience of cattle-breeders, stock-farmers, horse-dealers, and dog and poultry fanciers. Nay, it is only the other day that an eminent physiologist, Dr. Brown-Sequard, communicated to the Royal Society his discovery that epilepsy, artificially produced in guinea-pigs, by a means which he has discovered, is transmitted to their offspring.

But a race, once produced, is no more a fixed and immutable entity than the stock whence it sprang; variations arise among its members, and as these variations are transmitted like any others, new races may be developed out of the pre-existing one 'ad infinitum', or, at least, within any limit at present determined. Given sufficient time and sufficiently careful selection, and the multitude of races which may arise from a common stock is as astonishing as are the extreme structural differences which they may present. A remarkable example of this is to be found in the rock-pigeon, which Dr. Darwin has, in our opinion, satisfactorily demonstrated to be the progenitor of all our domestic pigeons, of which there are certainly more than a hundred well-marked races. The most noteworthy of these races are, the four great stocks known to the "fancy" as tumblers, pouters, carriers, and fantails; birds which not only differ most singularly in size, colour, and habits, but in the form of the beak and of the skull: in the proportions of the beak to the skull; in the number of tail-feathers; in the absolute and relative size of the feet; in the presence or absence of the uropygial gland; in the number of vertebrae in the back; in short, in precisely those characters in which the genera and species of birds differ from one another.

And it is most remarkable and instructive to observe, that none of these races can be shown to have been originated by the action of changes in what are commonly called external circumstances, upon the wild rock-pigeon. On the contrary, from time immemorial, pigeon-fanciers have had essentially similar methods of treating their pets, which have been housed, fed, protected and cared for in much the same way in all pigeonries. In fact, there is no case better adapted than that of the pigeons to refute the doctrine which one sees put forth on high authority, that "no other characters than those founded on the development of bone for the attachment of muscles" are capable of variation. In precise contradiction of this hasty assertion, Mr. Darwin's researches prove that the skeleton of the wings in domestic pigeons has hardly varied at all from that of the wild type; while, on the other hand, it is in exactly those respects, such as the relative length of the beak and skull, the number of the vertebrae, and the number of the tail-feathers, in which muscular exertion can have no important influence, that the utmost amount of variation has taken place.

We have said that the following out of the properties exhibited by physiological species would lead us into difficulties, and at this point they begin to be obvious; for if, as the result of spontaneous variation and of selective breeding, the progeny of a common stock may become separated into groups distinguished from one another by constant, not sexual, morphological characters, it is clear that the physiological definition of species is likely to clash with the morphological definition. No one would hesitate to describe the pouter and the tumbler as distinct species, if they were found fossil, or if their skins and skeletons were imported, as those of exotic wild birds commonly are—and without doubt, if considered alone, they are good and distinct morphological species. On the other hand, they are not physiological species, for they are descended from a common stock, the rock-pigeon.

Under these circumstances, as it is admitted on all sides that races occur in Nature, how are we to know whether any apparently distinct animals are really of different physiological species, or not, seeing that the amount of morphological difference is no safe guide? Is there any test of a physiological species? The usual answer of physiologists is in the affirmative. It is said that such a test is to be found in the phenomena of hybridization—in the results of crossing races, as compared with the results of crossing species.

So far as the evidence goes at present, individuals, of what are certainly known to be mere races produced by selection, however distinct they may appear to be, not only breed freely together, but the offspring of such crossed races are only perfectly fertile with one another. Thus, the spaniel and the greyhound, the dray-horse and the Arab, the pouter and the tumbler, breed together with perfect freedom, and their mongrels, if matched with other mongrels of the same kind, are equally fertile.

On the other hand, there can be no doubt that the individuals of many natural species are either absolutely infertile if crossed with individuals of other species, or, if they give rise to hybrid offspring, the hybrids so produced are infertile when paired together. The horse and the ass, for instance, if so crossed, give rise to the mule, and there is no certain evidence of offspring ever having been produced by a male and female mule. The unions of the rock-pigeon and the ring-pigeon appear to be equally barren of result. Here, then, says the physiologist, we have a means of distinguishing any two true species from any two varieties. If a male and a female, selected from each group, produce offspring, and that offspring is fertile with others produced in the same way, the groups are races and not species. If, on the other hand, no result ensues, or if the offspring are infertile with others produced in the same way, they are true physiological species. The test would be an admirable one, if, in the first place, it were always practicable to apply it, and if, in the second, it always yielded results susceptible of a definite interpretation. Unfortunately, in the great majority of cases, this touchstone for species is wholly inapplicable.

The constitution of many wild animals is so altered by confinement that they will not breed even with their own females, so that the negative results obtained from crosses are of no value; and the antipathy of wild animals of the same species for one another, or even of wild and tame members of the same species, is ordinarily so great, that it is hopeless to look for such unions in Nature. The hermaphrodism of most plants, the difficulty in the way of insuring the absence of their own, or the proper working of other pollen, are obstacles of no less magnitude in applying the test to them. And, in both animals and plants, is superadded the further difficulty, that experiments must be continued over a long time for the purpose of ascertaining the fertility of the mongrel or hybrid progeny, as well as of the first crosses from which they spring.

Not only do these great practical difficulties lie in the way of applying the hybridization test, but even when this oracle can be questioned, its replies are sometimes as doubtful as those of Delphi. For example, cases are cited by Mr. Darwin, of plants which are more fertile with the pollen of another species than with their own; and there are others, such as certain 'fuci', whose male element will fertilize the ovule of a plant of distinct species, while the males of the latter species are ineffective with the females of the first. So that, in the last-named instance, a physiologist, who should cross the two species in one way, would decide that they were true species; while another, who should cross them in the reverse way, would, with equal justice, according to the rule, pronounce them to be mere races. Several plants, which there is great reason to believe are mere varieties, are almost sterile when crossed; while both animals and plants, which have always been regarded by naturalists as of distinct species, turn out, when the test is applied, to be perfectly fertile. Again, the sterility or fertility of crosses seems to bear no relation to the structural resemblances or differences of the members of any two groups.

Mr. Darwin has discussed this question with singular ability and circumspection, and his conclusions are summed up as follows, at page 276 of his work:—

"First crosses between forms sufficiently distinct to be ranked as species, and their hybrids, are very generally, but not universally, sterile. The sterility is of all degrees, and is often so slight that the two most careful experimentalists who have ever lived have come to diametrically opposite conclusions in ranking forms by this test. The sterility is innately variable in individuals of the same species, and is eminently susceptible of favourable and unfavourable conditions. The degree of sterility does not strictly follow systematic affinity, but is governed by several curious and complex laws. It is generally different and sometimes widely different, in reciprocal crosses between the same two species. It is not always equal in degree in a first cross, and in the hybrid produced from this cross.

"In the same manner as in grafting trees, the capacity of one species or variety to take on another is incidental on generally unknown differences in their vegetative systems; so in crossing, the greater or less facility of one species to unite with another is incidental on unknown differences in their reproductive systems. There is no more reason to think that species have been specially endowed with various degrees of sterility to prevent them crossing and breeding in Nature, than to think that trees have been specially endowed with various and somewhat analogous degrees of difficulty in being grafted together, in order to prevent them becoming inarched in our forests.

"The sterility of first crosses between pure species, which have their reproductive systems perfect, seems to depend on several circumstances; in some cases largely on the early death of the embryo. The sterility of hybrids which have their reproductive systems imperfect, and which have had this system and their whole organization disturbed by being compounded of two distinct species, seems closely allied to that sterility which so frequently affects pure species when their natural conditions of life have been disturbed. This view is supported by a parallelism of another kind: namely, that the crossing of forms, only slightly different, is favourable to the vigour and fertility of the offspring; and that slight changes in the conditions of life are apparently favourable to the vigour and fertility of all organic beings. It is not surprising that the degree of difficulty in uniting two species, and the degree of sterility of their hybrid offspring, should generally correspond, though due to distinct causes; for both depend on the amount of difference of some kind between the species which are crossed. Nor is it surprising that the facility of effecting a first cross, the fertility of hybrids produced from it, and the capacity of being grafted together—though this latter capacity evidently depends on widely different circumstances—should all run to a certain extent parallel with the systematic affinity of the forms which are subjected to experiment; for systematic affinity attempts to express all kinds of resemblance between all species.

"First crosses between forms known to be varieties, or sufficiently alike to be considered as varieties, and their mongrel offspring, are very generally, but not quite universally, fertile. Nor is this nearly general and perfect fertility surprising, when we remember how liable we are to argue in a circle with respect to varieties in a state of Nature; and when we remember that the greater number of varieties have been produced under domestication by the selection of mere external differences, and not of differences in the reproductive system. In all other respects, excluding fertility, there is a close general resemblance between hybrids and mongrels."—Pp. 276-8.

We fully agree with the general tenor of this weighty passage; but forcible as are these arguments, and little as the value of fertility or infertility as a test of species may be, it must not be forgotten that the really important fact, so far as the inquiry into the origin of species goes, is, that there are such things in Nature as groups of animals and of plants, whose members are incapable of fertile union with those of other groups; and that there are such things as hybrids, which are absolutely sterile when crossed with other hybrids. For, if such phenomena as these were exhibited by only two of those assemblages of living objects, to which the name of species (whether it be used in its physiological or in its morphological sense) is given, it would have to be accounted for by any theory of the origin of species, and every theory which could not account for it would be, so far, imperfect.

Up to this point, we have been dealing with matters of fact, and the statements which we have laid before the reader would, to the best of our knowledge, be admitted to contain a fair exposition of what is at present known respecting the essential properties of species, by all who have studied the question. And whatever may be his theoretical views, no naturalist will probably be disposed to demur to the following summary of that exposition:—

Living beings, whether animals or plants, are divisible into multitudes of distinctly definable kinds, which are morphological species. They are also divisible into groups of individuals, which breed freely together, tending to reproduce their like, and are physiological species. Normally resembling their parents, the offspring of members of these species are still liable to vary; and the variation may be perpetuated by selection, as a race, which race, in many cases, presents all the characteristics of a morphological species. But it is not as yet proved that a race ever exhibits, when crossed with another race of the same species, those phenomena of hybridization which are exhibited by many species when crossed with other species. On the other hand, not only is it not proved that all species give rise to hybrids infertile 'inter se', but there is much reason to believe that, in crossing, species exhibit every gradation from perfect sterility to perfect fertility.

Such are the most essential characteristics of species. Even were man not one of them—a member of the same system and subject to the same laws—the question of their origin, their causal connexion, that is, with the other phenomena of the universe, must have attracted his attention, as soon as his intelligence had raised itself above the level of his daily wants.

Indeed history relates that such was the case, and has embalmed for us the speculations upon the origin of living beings, which were among the earliest products of the dawning intellectual activity of man. In those early days positive knowledge was not to be had, but the craving after it needed, at all hazards, to be satisfied, and according to the country, or the turn of thought, of the speculator, the suggestion that all living things arose from the mud of the Nile, from a primeval egg, or from some more anthropomorphic agency, afforded a sufficient resting-place for his curiosity. The myths of Paganism are as dead as Osiris or Zeus, and the man who should revive them, in opposition to the knowledge of our time, would be justly laughed to scorn; but the coeval imaginations current among the rude inhabitants of Palestine, recorded by writers whose very name and age are admitted by every scholar to be unknown, have unfortunately not yet shared their fate, but, even at this day, are regarded by nine-tenths of the civilized world as the authoritative standard of fact and the criterion of the justice of scientific conclusions, in all that relates to the origin of things, and, among them, of species. In this nineteenth century, as at the dawn of modern physical science, the cosmogony of the semi-barbarous Hebrew is the incubus of the philosopher and the opprobrium of the orthodox. Who shall number the patient and earnest seekers after truth, from the days of Galileo until now, whose lives have been embittered and their good name blasted by the mistaken zeal of Bibliolaters? Who shall count the host of weaker men whose sense of truth has been destroyed in the effort to harmonize impossibilities—whose life has been wasted in the attempt to force the generous new wine of Science into the old bottles of Judaism, compelled by the outcry of the same strong party?

It is true that if philosophers have suffered, their cause has been amply avenged. Extinguished theologians lie about the cradle of every science as the strangled snakes beside that of Hercules; and history records that whenever science and orthodoxy have been fairly opposed, the latter has been forced to retire from the lists, bleeding and crushed if not annihilated; scotched, if not slain. But orthodoxy is the Bourbon of the world of thought. It learns not, neither can it forget; and though, at present, bewildered and afraid to move, it is as willing as ever to insist that the first chapter of Genesis contains the beginning and the end of sound science; and to visit, with such petty thunderbolts as its half-paralysed hands can hurl, those who refuse to degrade Nature to the level of primitive Judaism.

Philosophers, on the other hand, have no such aggressive tendencies. With eyes fixed on the noble goal to which "per aspera et ardua" they tend, they may, now and then, be stirred to momentary wrath by the unnecessary obstacles with which the ignorant, or the malicious, encumber, if they cannot bar, the difficult path; but why should their souls be deeply vexed? The majesty of Fact is on their side, and the elemental forces of Nature are working for them. Not a star comes to the meridian at its calculated time but testifies to the justice of their methods—their beliefs are "one with falling rain and with the growing corn." By doubt they are established, and open inquiry is their bosom friend. Such men have no fear of traditions however venerable, and no respect for them when they become mischievous and obstructive; but they have better than mere antiquarian business in hand, and if dogmas, which ought to be fossil but are not, are not forced upon their notice, they are too happy to treat them as non-existent.

The hypotheses respecting the origin of species which profess to stand upon a scientific basis, and, as such, alone demand serious attention, are of two kinds. The one, the "special creation" hypothesis, presumes every species to have originated from one or more stocks, these not being the result of the modification of any other form of living matter—or arising by natural agencies—but being produced, as such, by a supernatural creative act.

The other, the so-called "transmutation" hypothesis, considers that all existing species are the result of the modification of pre-existing species, and those of their predecessors, by agencies similar to those which at the present day produce varieties and races, and therefore in an altogether natural way; and it is a probable, though not a necessary consequence of this hypothesis, that all living beings have arisen from a single stock. With respect to the origin of this primitive stock, or stocks, the doctrine of the origin of species is obviously not necessarily concerned. The transmutation hypothesis, for example, is perfectly consistent either with the conception of a special creation of the primitive germ, or with the supposition of its having arisen, as a modification of inorganic matter, by natural causes.

The doctrine of special creation owes its existence very largely to the supposed necessity of making science accord with the Hebrew cosmogony; but it is curious to observe that, as the doctrine is at present maintained by men of science, it is as hopelessly inconsistent with the Hebrew view as any other hypothesis.

If there be any result which has come more clearly out of geological investigation than another, it is, that the vast series of extinct animals and plants is not divisible, as it was once supposed to be, into distinct groups, separated by sharply-marked boundaries. There are no great gulfs between epochs and formations—no successive periods marked by the appearance of plants, of water animals, and of land animals, 'en masse'. Every year adds to the list of links between what the older geologists supposed to be widely separated epochs: witness the crags linking the drift with older tertiaries; the Maestricht beds linking the tertiaries with the chalk; the St. Cassian beds exhibiting an abundant fauna of mixed mesozoic and palaeozoic types, in rocks of an epoch once supposed to be eminently poor in life; witness, lastly, the incessant disputes as to whether a given stratum shall be reckoned devonian or carboniferous, silurian or devonian, cambrian or silurian.

This truth is further illustrated in a most interesting manner by the impartial and highly competent testimony of M. Pictet, from whose calculations of what percentage of the genera of animals, existing in any formation, lived during the preceding formation, it results that in no case is the proportion less than 'one-third', or 33 per cent. It is the triassic formation, or the commencement of the mesozoic epoch, which has received the smallest inheritance from preceding ages. The other formations not uncommonly exhibit 60, 80, or even 94 per cent. of genera in common with those whose remains are imbedded in their predecessor. Not only is this true, but the subdivisions of each formation exhibit new species characteristic of, and found only in, them; and, in many cases, as in the lias for example, the separate beds of these subdivisions are distinguished by well-marked and peculiar forms of life. A section, a hundred feet thick, will exhibit, at different heights, a dozen species of ammonite, none of which passes beyond its particular zone of limestone, or clay, into the zone below it or into that above it; so that those who adopt the doctrine of special creation must be prepared to admit, that at intervals of time, corresponding with the thickness of these beds, the Creator thought fit to interfere with the natural course of events for the purpose of making a new ammonite. It is not easy to transplant oneself into the frame of mind of those who can accept such a conclusion as this, on any evidence short of absolute demonstration; and it is difficult to see what is to be gained by so doing, since, as we have said, it is obvious that such a view of the origin of living beings is utterly opposed to the Hebrew cosmogony. Deserving no aid from the powerful arm of Bibliolatry, then, does the received form of the hypothesis of special creation derive any support from science or sound logic? Assuredly not much. The arguments brought forward in its favour all take one form: If species were not supernaturally created, we cannot understand the facts 'x' or 'y', or 'z'; we cannot understand the structure of animals or plants, unless we suppose they were contrived for special ends; we cannot understand the structure of the eye, except by supposing it to have been made to see with; we cannot understand instincts, unless we suppose animals to have been miraculously endowed with them.

As a question of dialectics, it must be admitted that this sort of reasoning is not very formidable to those who are not to be frightened by consequences. It is an 'argumentum ad ignorantiam'—take this explanation or be ignorant.

But suppose we prefer to admit our ignorance rather than adopt a hypothesis at variance with all the teachings of Nature? Or, suppose for a moment we admit the explanation, and then seriously ask ourselves how much the wiser are we; what does the explanation explain? Is it any more than a grandiloquent way of announcing the fact, that we really know nothing about the matter? A phenomenon is explained when it is shown to be a case of some general law of Nature; but the supernatural interposition of the Creator can, by the nature of the case, exemplify no law, and if species have really arisen in this way, it is absurd to attempt to discuss their origin.

Or, lastly, let us ask ourselves whether any amount of evidence which the nature of our faculties permits us to attain, can justify us in asserting that any phenomenon is out of the reach of natural causation. To this end it is obviously necessary that we should know all the consequences to which all possible combinations, continued through unlimited time, can give rise. If we knew these, and found none competent to originate species, we should have good ground for denying their origin by natural causation. Till we know them, any hypothesis is better than one which involves us in such miserable presumption.

But the hypothesis of special creation is not only a mere specious mask for our ignorance; its existence in Biology marks the youth and imperfection of the science. For what is the history of every science but the history of the elimination of the notion of creative, or other interferences, with the natural order of the phenomena which are the subject-matter of that science? When Astronomy was young "the morning stars sang together for joy," and the planets were guided in their courses by celestial hands. Now, the harmony of the stars has resolved itself into gravitation according to the inverse squares of the distances, and the orbits of the planets are deducible from the laws of the forces which allow a schoolboy's stone to break a window. The lightning was the angel of the Lord; but it has pleased Providence, in these modern times, that science should make it the humble messenger of man, and we know that every flash that shimmers about the horizon on a summer's evening is determined by ascertainable conditions, and that its direction and brightness might, if our knowledge of these were great enough, have been calculated.

The solvency of great mercantile companies rests on the validity of the laws which have been ascertained to govern the seeming irregularity of that human life which the moralist bewails as the most uncertain of things; plague, pestilence, and famine are admitted, by all but fools, to be the natural result of causes for the most part fully within human control, and not the unavoidable tortures inflicted by wrathful Omnipotence upon His helpless handiwork.

Harmonious order governing eternally continuous progress—the web and woof of matter and force interweaving by slow degrees, without a broken thread, that veil which lies between us and the Infinite—that universe which alone we know or can know; such is the picture which science draws of the world, and in proportion as any part of that picture is in unison with the rest, so may we feel sure that it is rightly painted. Shall Biology alone remain out of harmony with her sister sciences?

Such arguments against the hypothesis of the direct creation of species as these are plainly enough deducible from general considerations; but there are, in addition, phenomena exhibited by species themselves, and yet not so much a part of their very essence as to have required earlier mention, which are in the highest degree perplexing, if we adopt the popularly accepted hypothesis. Such are the facts of distribution in space and in time; the singular phenomena brought to light by the study of development; the structural relations of species upon which our systems of classification are founded; the great doctrines of philosophical anatomy, such as that of homology, or of the community of structural plan exhibited by large groups of species differing very widely in their habits and functions.

The species of animals which inhabit the sea on opposite sides of the isthmus of Panama are wholly distinct [4] the animals and plants which inhabit islands are commonly distinct from those of the neighbouring mainlands, and yet have a similarity of aspect.

The mammals of the latest tertiary epoch in the Old and New Worlds belong to the same genera, or family groups, as those which now inhabit the same great geographical area. The crocodilian reptiles which existed in the earliest secondary epoch were similar in general structure to those now living, but exhibit slight differences in their vertebrae, nasal passages, and one or two other points. The guinea-pig has teeth which are shed before it is born, and hence can never subserve the masticatory purpose for which they seem contrived, and, in like manner, the female dugong has tusks which never cut the gum. All the members of the same great group run through similar conditions in their development, and all their parts, in the adult state, are arranged according to the same plan. Man is more like a gorilla than a gorilla is like a lemur. Such are a few, taken at random, among the multitudes of similar facts which modern research has established; but when the student seeks for an explanation of them from the supporters of the received hypothesis of the origin of species, the reply he receives is, in substance, of Oriental simplicity and brevity—"Mashallah! it so pleases God!" There are different species on opposite sides of the isthmus of Panama, because they were created different on the two sides. The pliocene mammals are like the existing ones, because such was the plan of creation; and we find rudimental organs and similarity of plan, because it has pleased the Creator to set before Himself a "divine exemplar or archetype," and to copy it in His works; and somewhat ill, those who hold this view imply, in some of them. That such verbal hocus-pocus should be received as science will one day be regarded as evidence of the low state of intelligence in the nineteenth century, just as we amuse ourselves with the phraseology about Nature's abhorrence of a vacuum, wherewith Torricelli's compatriots were satisfied to explain the rise of water in a pump. And be it recollected that this sort of satisfaction works not only negative but positive ill, by discouraging inquiry, and so depriving man of the usufruct of one of the most fertile fields of his great patrimony, Nature.

The objections to the doctrine of the origin of species by special creation which have been detailed, must have occurred, with more or less force, to the mind of every one who has seriously and independently considered the subject. It is therefore no wonder that, from time to time, this hypothesis should have been met by counter hypotheses, all as well, and some better founded than itself; and it is curious to remark that the inventors of the opposing views seem to have been led into them as much by their knowledge of geology, as by their acquaintance with biology. In fact, when the mind has once admitted the conception of the gradual production of the present physical state of our globe, by natural causes operating through long ages of time, it will be little disposed to allow that living beings have made their appearance in another way, and the speculations of De Maillet and his successors are the natural complement of Scilla's demonstration of the true nature of fossils.

A contemporary of Newton and of Leibnitz, sharing therefore in the intellectual activity of the remarkable age which witnessed the birth of modern physical science, Benoit de Maillet spent a long life as a consular agent of the French Government in various Mediterranean ports. For sixteen years, in fact, he held the office of Consul-General in Egypt, and the wonderful phenomena offered by the valley of the Nile appear to have strongly impressed his mind, to have directed his attention to all facts of a similar order which came within his observation, and to have led him to speculate on the origin of the present condition of our globe and of its inhabitants. But, with all his ardour for science, De Maillet seems to have hesitated to publish views which, notwithstanding the ingenious attempts to reconcile them with the Hebrew hypothesis contained in the preface to "Telliamed," were hardly likely to be received with favour by his contemporaries.

But a short time had elapsed since more than one of the great anatomists and physicists of the Italian school had paid dearly for their endeavours to dissipate some of the prevalent errors; and their illustrious pupil, Harvey, the founder of modern physiology, had not fared so well, in a country less oppressed by the benumbing influences of theology, as to tempt any man to follow his example. Probably not uninfluenced by these considerations, his Catholic majesty's Consul-General for Egypt kept his theories to himself throughout a long life, for 'Telliamed,' the only scientific work which is known to have proceeded from his pen, was not printed till 1735, when its author had reached the ripe age of seventy-nine; and though De Maillet lived three years longer, his book was not given to the world before 1748. Even then it was anonymous to those who were not in the secret of the anagrammatic character of its title; and the preface and dedication are so worded as, in case of necessity, to give the printer a fair chance of falling back on the excuse that the work was intended for a mere 'jeu d'esprit'.

The speculations of the suppositious Indian sage, though quite as sound as those of many a "Mosaic Geology," which sells exceedingly well, have no great value if we consider them by the light of modern science. The waters are supposed to have originally covered the whole globe; to have deposited the rocky masses which compose its mountains by processes comparable to those which are now forming mud, sand, and shingle; and then to have gradually lowered their level, leaving the spoils of their animal and vegetable inhabitants embedded in the strata. As the dry land appeared, certain of the aquatic animals are supposed to have taken to it, and to have become gradually adapted to terrestrial and aerial modes of existence. But if we regard the general tenor and style of the reasoning in relation to the state of knowledge of the day, two circumstances appear very well worthy of remark. The first, that De Maillet had a notion of the modifiability of living forms (though without any precise information on the subject), and how such modifiability might account for the origin of species; the second, that he very clearly apprehended the great modern geological doctrine, so strongly insisted upon by Hutton, and so ably and comprehensively expounded by Lyell, that we must look to existing causes for the explanation of past geological events. Indeed, the following passage of the preface, in which De Maillet is supposed to speak of the Indian philosopher Telliamed, his 'alter ego', might have been written by the most philosophical uniformitarian of the present day:—

"Ce qu'il y a d'etonnant, est que pour arriver a ces connoissances il semble avoir perverti l'ordre naturel, puisqu'au lieu de s'attacher d'abord a rechercher l'origine de notre globe il a commence par travailler a s'instruire de la nature. Mais a l'entendre, ce renversement de l'ordre a ete pour lui l'effet d'un genie favorable qui l'a conduit pas a pas et comme par la main aux decouvertes les plus sublimes. C'est en decomposant la substance de ce globe par une anatomie exacte de toutes ses parties qu'il a premierement appris de quelles matieres il etait compose et quels arrangemens ces memes matieres observaient entre elles. Ces lumieres jointes a l'esprit de comparaison toujours necessaire a quiconque entreprend de percer les voiles dont la nature aime a se cacher, ont servi de guide a notre philosophe pour parvenir a des connoissances plus interessantes. Par la matiere et l'arrangement de ces compositions il pretend avoir reconnu quelle est la veritable origine de ce globe que nous habitons, comment et par qui il a ete forme."—Pp. xix. xx.

But De Maillet was before his age, and as could hardly fail to happen to one who speculated on a zoological and botanical question before Linnaeus, and on a physiological problem before Haller, he fell into great errors here and there; and hence, perhaps, the general neglect of his work. Robinet's speculations are rather behind, than in advance of, those of De Maillet; and though Linnaeus may have played with the hypothesis of transmutation, it obtained no serious support until Lamarck adopted it, and advocated it with great ability in his 'Philosophie Zoologique.'

Impelled towards the hypothesis of the transmutation of species, partly by his general cosmological and geological views; partly by the conception of a graduated, though irregularly branching, scale of being, which had arisen out of his profound study of plants and of the lower forms of animal life, Lamarck, whose general line of thought often closely resembles that of De Maillet, made a great advance upon the crude and merely speculative manner in which that writer deals with the question of the origin of living beings, by endeavouring to find physical causes competent to effect that change of one species into another, which De Maillet had only supposed to occur. And Lamarck conceived that he had found in Nature such causes, amply sufficient for the purpose in view. It is a physiological fact, he says, that organs are increased in size by action, atrophied by inaction; it is another physiological fact that modifications produced are transmissible to offspring. Change the actions of an animal, therefore, and you will change its structure, by increasing the development of the parts newly brought into use and by the diminution of those less used; but by altering the circumstances which surround it you will alter its actions, and hence, in the long run, change of circumstance must produce change of organization. All the species of animals, therefore, are, in Lamarck's view, the result of the indirect action of changes of circumstance, upon those primitive germs which he considered to have originally arisen, by spontaneous generation, within the waters of the globe. It is curious, however, that Lamarck should insist so strongly [5] as he has done, that circumstances never in any degree directly modify the form or the organization of animals, but only operate by changing their wants and consequently their actions; for he thereby brings upon himself the obvious question, how, then, do plants, which cannot be said to have wants or actions, become modified? To this he replies, that they are modified by the changes in their nutritive processes, which are effected by changing circumstances; and it does not seem to have occurred to him that such changes might be as well supposed to take place among animals.

When we have said that Lamarck felt that mere speculation was not the way to arrive at the origin of species, but that it was necessary, in order to the establishment of any sound theory on the subject, to discover by observation or otherwise, some 'vera causa', competent to give rise to them; that he affirmed the true order of classification to coincide with the order of their development one from another; that he insisted on the necessity of allowing sufficient time, very strongly; and that all the varieties of instinct and reason were traced back by him to the same cause as that which has given rise to species, we have enumerated his chief contributions to the advance of the question. On the other hand, from his ignorance of any power in Nature competent to modify the structure of animals, except the development of parts, or atrophy of them, in consequence of a change of needs, Lamarck was led to attach infinitely greater weight than it deserves to this agency, and the absurdities into which he was led have met with deserved condemnation. Of the struggle for existence, on which, as we shall see, Mr. Darwin lays such great stress, he had no conception; indeed, he doubts whether there really are such things as extinct species, unless they be such large animals as may have met their death at the hands of man; and so little does he dream of there being any other destructive causes at work, that, in discussing the possible existence of fossil shells, he asks, "Pourquoi d'ailleurs seroient-ils perdues des que l'homme n'a pu operer leur destruction?" ('Phil. Zool.,' vol. i. p. 77.) Of the influence of selection Lamarck has as little notion, and he makes no use of the wonderful phenomena which are exhibited by domesticated animals, and illustrate its powers. The vast influence of Cuvier was employed against the Lamarckian views, and, as the untenability of some of his conclusions was easily shown, his doctrines sank under the opprobrium of scientific, as well as of theological, heterodoxy. Nor have the efforts made of late years to revive them tended to re-establish their credit in the minds of sound thinkers acquainted with the facts of the case; indeed it may be doubted whether Lamarck has not suffered more from his friends than from his foes.

Two years ago, in fact, though we venture to question if even the strongest supporters of the special creation hypothesis had not, now and then, an uneasy consciousness that all was not right, their position seemed more impregnable than ever, if not by its own inherent strength, at any rate by the obvious failure of all the attempts which had been made to carry it. On the other hand, however much the few, who thought deeply on the question of species, might be repelled by the generally received dogmas, they saw no way of escaping from them save by the adoption of suppositions so little justified by experiment or by observation as to be at least equally distasteful.

The choice lay between two absurdities and a middle condition of uneasy scepticism; which last, however unpleasant and unsatisfactory, was obviously the only justifiable state of mind under the circumstances.

Such being the general ferment in the minds of naturalists, it is no wonder that they mustered strong in the rooms of the Linnaean Society, on the 1st of July of the year 1858, to hear two papers by authors living on opposite sides of the globe, working out their results independently, and yet professing to have discovered one and the same solution of all the problems connected with species. The one of these authors was an able naturalist, Mr. Wallace, who had been employed for some years in studying the productions of the islands of the Indian Archipelago, and who had forwarded a memoir embodying his views to Mr. Darwin, for communication to the Linnaean Society. On perusing the essay, Mr. Darwin was not a little surprised to find that it embodied some of the leading ideas of a great work which he had been preparing for twenty years, and parts of which, containing a development of the very same views, had been perused by his private friends fifteen or sixteen years before. Perplexed in what manner to do full justice both to his friend and to himself, Mr. Darwin placed the matter in the hands of Dr. Hooker and Sir Charles Lyell, by whose advice he communicated a brief abstract of his own views to the Linnaean Society, at the same time that Mr. Wallace's paper was read. Of that abstract, the work on the 'Origin of Species' is an enlargement; but a complete statement of Mr. Darwin's doctrine is looked for in the large and well-illustrated work which he is said to be preparing for publication.

The Darwinian hypothesis has the merit of being eminently simple and comprehensible in principle, and its essential positions may be stated in a very few words: all species have been produced by the development of varieties from common stocks; by the conversion of these, first into permanent races and then into new species, by the process of 'natural selection', which process is essentially identical with that artificial selection by which man has originated the races of domestic animals—the 'struggle for existence' taking the place of man, and exerting, in the case of natural selection, that selective action which he performs in artificial selection.

The evidence brought forward by Mr. Darwin in support of his hypothesis is of three kinds. First, he endeavours to prove that species may be originated by selection; secondly, he attempts to show that natural causes are competent to exert selection; and thirdly, he tries to prove that the most remarkable and apparently anomalous phenomena exhibited by the distribution, development, and mutual relations of species, can be shown to be deducible from the general doctrine of their origin, which he propounds, combined with the known facts of geological change; and that, even if all these phenomena are not at present explicable by it, none are necessarily inconsistent with it.

There cannot be a doubt that the method of inquiry which Mr. Darwin has adopted is not only rigorously in accordance with the canons of scientific logic, but that it is the only adequate method. Critics exclusively trained in classics or in mathematics, who have never determined a scientific fact in their lives by induction from experiment or observation, prate learnedly about Mr. Darwin's method, which is not inductive enough, not Baconian enough, forsooth, for them. But even if practical acquaintance with the process of scientific investigation is denied them, they may learn, by the perusal of Mr. Mill's admirable chapter "On the

Deductive Method," that there are multitudes of scientific inquiries in which the method of pure induction helps the investigator but a very little way.

"The mode of investigation," says Mr. Mill, "which, from the proved inapplicability of direct methods of observation and experiment, remains to us as the main source of the knowledge we possess, or can acquire, respecting the conditions and laws of recurrence of the more complex phenomena, is called, in its most general expression, the deductive method, and consists of three operations: the first, one of direct induction; the second, of ratiocination; and the third, of verification."

Now, the conditions which have determined the existence of species are not only exceedingly complex, but, so far as the great majority of them are concerned, are necessarily beyond our cognizance. But what Mr. Darwin has attempted to do is in exact accordance with the rule laid down by Mr. Mill; he has endeavoured to determine certain great facts inductively, by observation and experiment; he has then reasoned from the data thus furnished; and lastly, he has tested the validity of his ratiocination by comparing his deductions with the observed facts of Nature. Inductively, Mr. Darwin endeavours to prove that species arise in a given way. Deductively, he desires to show that, if they arise in that way, the facts of distribution, development, classification, etc., may be accounted for, 'i.e.' may be deduced from their mode of origin, combined with admitted changes in physical geography and climate, during an indefinite period. And this explanation, or coincidence of observed with deduced facts, is, so far as it extends, a verification of the Darwinian view.

There is no fault to be found with Mr. Darwin's method, then; but it is another question whether he has fulfilled all the conditions imposed by that method. Is it satisfactorily proved, in fact, that species may be originated by selection? that there is such a thing as natural selection? that none of the phenomena exhibited by species are inconsistent with the origin of species in this way? If these questions can be answered in the affirmative, Mr. Darwin's view steps out of the rank of hypotheses into those of proved theories; but, so long as the evidence at present adduced falls short of enforcing that affirmation, so long, to our minds, must the new doctrine be content to remain among the former—an extremely valuable, and in the highest degree probable, doctrine, indeed the only extant hypothesis which is worth anything in a scientific point of view; but still a hypothesis, and not yet the theory of species.

After much consideration, and with assuredly no bias against Mr. Darwin's views, it is our clear conviction that, as the evidence stands, it is not absolutely proven that a group of animals, having all the characters exhibited by species in Nature, has ever been originate by selection, whether artificial or natural. Groups having the morphological character of species, distinct and permanent races in fact, have been so produced over and over again; but there is no positive evidence, at present, that any group of animals has, by variation and selective breeding, given rise to another group which was, even in the least degree, infertile with the first. Mr. Darwin is perfectly aware of this weak point, and brings forward a multitude of ingenious and important arguments to diminish the force of the objection. We admit the value of these arguments to their fullest extent; nay, we will go so far as to express our belief that experiments, conducted by a skilful physiologist, would very probably obtain the desired production of mutually more or less infertile breeds from a common stock, in a comparatively few years; but still, as the case stands at present, this "little rift within the lute" is not to be disguised nor overlooked.

In the remainder of Mr. Darwin's argument our own private ingenuity has not hitherto enabled us to pick holes of any great importance; and judging by what we hear and read, other adventurers in the same field do not seem to have been much more fortunate. It has been urged, for instance, that in his chapters on the struggle for existence and on natural selection, Mr. Darwin does not so much prove that natural selection does occur, as that it must occur; but, in fact, no other sort of demonstration is attainable. A race does not attract our attention in Nature until it has, in all probability, existed for a considerable time, and then it is too late to inquire into the conditions of its origin. Again, it is said that there is no real analogy between the selection which takes place under domestication, by human influence, and any operation which can be effected by Nature, for man interferes intelligently. Reduced to its elements, this argument implies that an effect produced with trouble by an intelligent agent must, 'a fortiori', be more troublesome, if not impossible, to an unintelligent agent. Even putting aside the question whether Nature, acting as she does according to definite and invariable laws, can be rightly called an unintelligent agent, such a position as this is wholly untenable. Mix salt and sand, and it shall puzzle the wisest of men, with his mere natural appliances, to separate all the grains of sand from all the grains of salt; but a shower of rain will effect the same object in ten minutes. And so, while man may find it tax all his intelligence to separate any variety which arises, and to breed selectively from it, the destructive agencies incessantly at work in Nature, if they find one variety to be more soluble in circumstances than the other, will inevitably, in the long run, eliminate it.

A frequent and a just objection to the Lamarckian hypothesis of the transmutation of species is based upon the absence of transitional forms between many species. But against the Darwinian hypothesis this argument has no force. Indeed, one of the most valuable and suggestive parts of Mr. Darwin's work is that in which he proves, that the frequent absence of transitions is a necessary consequence of his doctrine, and that the stock whence two or more species have sprung, need in no respect be intermediate between these species. If any two species have arisen from a common stock in the same way as the carrier and the pouter, say, have arisen from the rock-pigeon, then the common stock of these two species need be no more intermediate between the two than the rock-pigeon is between the carrier and pouter. Clearly appreciate the force of this analogy, and all the arguments against the origin of species by selection, based on the absence of transitional forms, fall to the ground. And Mr. Darwin's position might, we think, have been even stronger than it is if he had not embarrassed himself with the aphorism, "Natura non facit saltum," which turns up so often in his pages. We believe, as we have said above, that Nature does make jumps now and then, and a recognition of the fact is of no small importance in disposing of many minor objections to the doctrine of transmutation.

But we must pause. The discussion of Mr. Darwin's arguments in detail would lead us far beyond the limits within which we proposed, at starting, to confine this article. Our object has been attained if we have given an intelligible, however brief, account of the established facts connected with species, and of the relation of the explanation of those facts offered by Mr. Darwin to the theoretical views held by his predecessors and his contemporaries, and, above all, to the requirements of scientific logic. We have ventured to point out that it does not, as yet, satisfy all those requirements; but we do not hesitate to assert that it is as superior to any preceding or contemporary hypothesis, in the extent of observational and experimental basis on which it rests, in its rigorously scientific method, and in its power of explaining biological phenomena, as was the hypothesis of Copernicus to the speculations of Ptolemy. But the planetary orbits turned out to be not quite circular after all, and, grand as was the service Copernicus rendered to science, Kepler and Newton had to come after him. What if the orbit of Darwinism should be a little too circular? What if species should offer residual phenomena, here and there, not explicable by natural selection? Twenty years hence naturalists may be in a position to say whether this is, or is not, the case; but in either event they will owe the author of 'The Origin of Species' an immense debt of gratitude. We should leave a very wrong impression on the reader's mind if we permitted him to suppose that the value of that work depends wholly on the ultimate justification of the theoretical views which it contains. On the contrary, if they were disproved to-morrow, the book would still be the best of its kind—the most compendious statement of well-sifted facts bearing on the doctrine of species that has ever appeared. The chapters on Variation, on the Struggle for Existence, on Instinct, on Hybridism, on the Imperfection of the Geological Record, on Geographical Distribution, have not only no equals, but, so far as our knowledge goes, no competitors, within the range of biological literature. And viewed as a whole, we do not believe that, since the publication of Von Baer's Researches on Development, thirty years ago, any work has appeared calculated to exert so large an influence, not only on the future of Biology, but in extending the domination of Science over regions of thought into which she has, as yet, hardly penetrated.

Naturally Selected — Thomas H. Huxley

Thomas Henry Huxley was one of the first adherents to Darwin's theory of evolution by natural selection, and did more than anyone else to advance its acceptance among scientists and the public alike. As is evident from the letter quoted above, Huxley was a passionate defender of Darwin's theory -- so passionate that he has been called "Darwin's Bulldog". But Huxley was not only the bulldog for Darwin's theory, but was a great biologist in his own right, who did original research in zoology and paleontology. Nor did he slavishly and uncritically swallow Darwin's theory; he criticized several aspects of it, pointing out a number of problems.

Biography of Huxley

He was born on May 4, 1825, in Ealing, near London, the seventh of eight children in a family that was none too affluent. Huxley's only childhood education was two years at Ealing school, where his father taught mathematics; this ended in 1835 when the family moved to Coventry. Despite his lack of formal education, young Huxley read voraciously in science, history, and philosophy, and taught himself German. At the age of 15, Huxley began a medical apprenticeship; soon he won a scholarship to study at Charing Cross Hospital. At 21, Huxley signed on as assistant surgeon on the H.M.S. *Rattlesnake*, a Royal Navy frigate assigned to chart the seas around Australia and New Guinea. Huxley vividly described conditions on the ship in his diary:

I wonder if it is possible for the mind of man to conceive anything more degradingly offensive than the condition of us 150 men, shut up in this wooden box, being watered with hot water, as we are now. . . It's too hot to sleep, and my sole amusement consists in watching the cockroaches, which are in a state of intense excitement and happiness.

Despite the excited cockroaches and the poor science facilities on board, Huxley collected and studied marine invertebrates, in particular cnidarians, tunicates, and cephalopod molluscs. (He also collected a fiancée, Henrietta Heathorn, whom he met and immediately fell in love with while in port in Sydney, Australia.) When he returned to England in October 1850, he found that his research results, which he had mailed back to England from each port of call, had won him acceptance into the ranks of the English scientific establishment. Huxley soon became acquainted with scientists like the geologist Charles Lyell, the botanist Joseph Hooker, the philosopher Herbert Spencer, and the naturalist Charles Darwin. Professional positions in science were rare at the time -- most naturalists were affluent amateurs -- but Huxley managed to support himself on a stipend from the Navy and by writing popular science articles. After leaving the Navy in 1854, Huxley managed to secure a lectureship at the School of Mines in London, and sent for his fiancée. They were married in 1855.

Huxley's Scientific Thought

As the nickname "Darwin's bulldog" would suggest, Huxley was an outspoken defender and advocate for Darwin's theory of evolution by natural selection. Perhaps surprisingly, he was at first an opponent of any evolutionary change at all, believing that the living world had stayed much the same for as far back as its history could be traced, and that modern taxa would eventually be found in the oldest rocks. But he came to accept evolutionary views: his reaction to reading the *Origin of Species* was "How stupid of me not to have thought of that."

He is best known for his famous debate in June 1860, at the British Association meeting at Oxford. His opponent, Archbishop Samuel Wilberforce, was not-so-affectionately known as "Soapy Sam" for his renowned slipperiness in debate. Wilberforce was coached against Huxley by Richard Owen. During the debate, Archbishop Wilberforce ridiculed evolution and asked Huxley whether he was descended from an ape on his grandmother's side or his grandfather's. Accounts vary as to exactly what happened next, but according to one telling of the story, Huxley muttered "The Lord hath delivered him into my hands," and then rose to give a brilliant defense of Darwin's theory, concluding with the rejoinder, "I would rather be the offspring of two apes than be a man and afraid to face the truth." Huxley's own retelling of the tale was a little different, and quite a bit less dramatic:

If then, said I, the question is put to me would I rather have a miserable ape for a grandfather or a man highly endowed by nature and possessed of great means of influence & yet who employs these faculties & that influence for the mere purpose of introducing ridicule into a grave scientific discussion, I unhesitatingly affirm my preference for the ape.

All accounts agree that Huxley trounced Wilberforce in the debate, defending evolution as the best explanation yet advanced for species diversity.

However, Huxley did not blindly follow Darwin's theory, and critiqued it even as he was defending it. In particular, where Darwin had seen evolution and a slow, gradual, continuous process, Huxley thought that an evolving lineage might make rapid jumps, or saltations. As he wrote to Darwin just before publication of the *Origin of Species*, "You have loaded yourself with an unnecessary difficulty in adopting *Natura non facit saltum* [Nature does not make leaps] so unreservedly."

Huxley's support for natural selection is perhaps surprising when contrasted with his earlier attacks on the evolutionary theories put forth by Lamarck and Robert Chambers. Both of these theories advocated some kind of progression -- some kind of general tendency present in all organisms to evolve "upward" into more and more complex forms. Huxley would have nothing to do with such progressionist ideas, which he regarded as being more metaphysical than scientific; this mistrust of progression lay behind his initial skepticism of all evolutionary ideas. Similarly, Huxley rejected the then-popular theory of recapitulation, following Karl von Baer (whose writings Huxley had translated from the German). Huxley wrote, "the progress of a higher animal in development is not through the forms of the lower, but through forms which are common to both lower and higher. . ."

Huxley's most famous writing, published in 1863, is *Evidence on Man's Place in Nature*. This book, published only five years after Darwin's *Origin of Species*, was a comprehensive review of what was known at the time about primate and human paleontology and ethology. More than that, it was the first attempt to apply evolution explicitly to the human race. Darwin had avoided direct mention of human evolution, stating only that "light will be thrown on the origin of Man;" Huxley explicitly presented evidence for human evolution. In this, once again, he locked horns with Richard Owen, who had claimed that the human brain contained parts that were not found in apes, and that therefore humans could not be classified with the apes nor descended from them. Huxley and his colleagues showed that the brains of apes and humans were fundamentally similar in every anatomical detail.

Huxley founded a remarkable dynasty of English scientists and thinkers. His son Leonard was a noted biographer and "man of letters." Leonard's oldest son Julian was one of the authors of the evolutionary synthesis of the early 20th century; Julian's son Francis became a noted anthropologist. Another of Leonard's sons, Andrew (later Sir Andrew Huxley) was also an eminent scientist, sharing the 1963 Nobel Prize in Physiology or Medicine for his work on nerve impulses and muscle contraction. Julian's brother Aldous Huxley was a novelist, screenwriter and essayist; his best-known book is the anti-utopia *Brave New World*. A more distant relative, Leonard George Golden Huxley, became a prominent physicist.

Excerpted from:

http://www.ucmp.berkeley.edu/history/thuxley.html

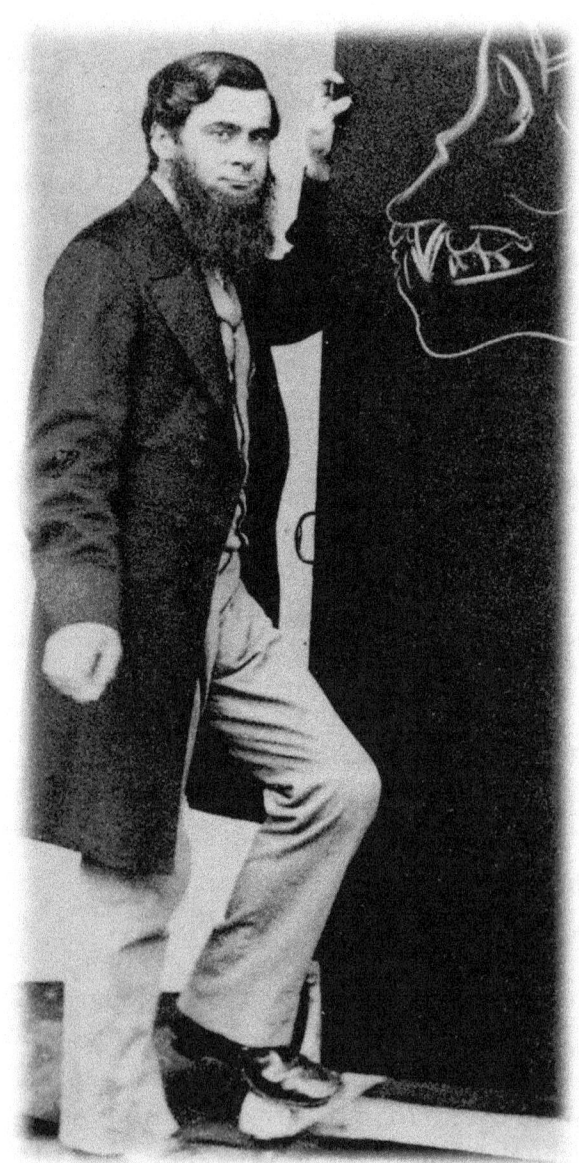

MSAC Philosophy Group Circular

| Mt. San Antonio College | Walnut, California |

Naturally Selected

Molecular Structure of Nucleic Acids

James Watson & Francis Crick

Nature, 171 (1953)

Original Papers in Biology
Issue Four

A Structure for Deoxyribose Nucleic Acid

J. D. Watson and F. H. C. Crick (1)

April 25, 1953 (2), *Nature* (3), 171, 737-738

We wish to suggest a structure for the salt of deoxyribose nucleic acid (D.N.A.). This structure has novel features which are of considerable biological interest.

A structure for nucleic acid has already been proposed by **Pauling (4)** and Corey[1]. They kindly made their manuscript available to us in advance of publication. Their model consists of three intertwined chains, with the phosphates near the fibre axis, and the bases on the outside. In our opinion, this structure is unsatisfactory for two reasons:

(1) We believe that the material which gives the X-ray diagrams is the salt, not the free acid. Without the acidic hydrogen atoms it is not clear what forces would hold the structure together, especially as the negatively charged phosphates near the axis will repel each other.

(2) Some of the van der Waals distances appear to be too small.

Another three-chain structure has also been suggested by Fraser (in the press). In his model the phosphates are on the outside and the bases on the inside, linked together by hydrogen bonds. This structure as described is rather ill-defined, and for this reason we shall not comment on it.

We wish to put forward a **radically different structure for the salt of deoxyribose nucleic acid (5)**. This structure has two helical chains each coiled round the same axis (see diagram). We have made the usual chemical assumptions, namely, that each chain consists of phosphate diester groups joining beta-D-deoxyribofuranose residues with 3',5' linkages. The two chains (but not their bases) are related by a dyad perpendicular to the fibre axis. Both chains follow right-handed helices, but owing to the dyad the sequences of the atoms in **the two chains run in opposite directions (6)**. Each chain loosely resembles **Furberg's**[2] **model No. 1 (7)**; that is, the bases are on the inside of the helix and the phosphates on the outside. The configuration of the sugar and the atoms near it is close to Furberg's "standard configuration," the sugar being roughly perpendicular to the attached base. There is a residue on each every 3.4 A. in the z-direction. We have assumed an angle of 36° between adjacent residues in the same chain, so that the structure repeats after 10 residues on each chain, that is, after 34 A. The distance of a phosphorus atom from the fibre axis is 10 A. As the phosphates are on the outside, cations have easy access to them.

Figure 1
This figure is purely diagrammatic (8). The two ribbons symbolize the two phophate-sugar chains, and the horizontal rods the pairs of bases holding the chains together. The vertical line marks the fibre axis.

The structure is an open one, and its water content is rather high. At lower water contents we would expect the bases to tilt so that the structure could become more compact.

The novel feature of the structure is the manner in which the two chains are held together by the purine and pyrimidine bases. The planes of the bases are perpendicular to the fibre axis. They are joined together in pairs, a single base from one chain being hydroden-bonded to a single base from the other chain, so that the two lie side by side with identical z-coordinates. One of the pair must be a purine and the other a pyrimidine for bonding to occur. The hydrogen bonds are made as follows: purine position 1 to pyrimidine position 1; purine position 6 to pyrimidine position 6.

If it is assumed that the bases only occur in the structure in the most plausible tautomeric forms (that is, with the keto rather than the enol configurations) it is found that only specific pairs of bases can bond together. **These pairs are: adenine (purine) with thymine (pyrimidine), and guanine (purine) with cytosine (pyrimidine) (9)**.

In other words, if an adenine forms one member of a pair, on either chain, then on these assumptions the other member must be thymine; similarly for guanine and cytosine. The sequence of bases on a single chain does not appear to be restricted in any way. However, if only specific pairs of bases can be formed, it follows that if the sequence of bases on one chain is given, then the sequence on the other chain is automatically determined.

It has been found experimentally (10)[3,4] that the ratio of the amounts of adenine to thymine, and the ratio of guanine to cytosine, are always very close to unity for deoxyribose nucleic acid.

It is probably impossible to build this structure with a ribose sugar in place of the deoxyribose, as the extra oxygen atom would make too close a van der Waals contact.

The previously published X-ray data[5,6] on deoxyribose nucleic acid are insufficient for a rigorous test of our structure. So far as we can tell, it is roughly compatible with the experimental data, but it must be regarded as unproved until it has been checked against more exact results. **Some of these are given in the following communications (11)**. We were **not aware of the details of the results presented there when we devised our structure (12)**, which rests mainly though not entirely on published experimental data and stereochemical arguments.

It has not escaped our notice (13) that the specific pairing we have postulated immediately suggests a possible copying mechanism for the genetic material.

Full details of the structure, including the conditions assumed in building it, together with a set of coordinates for the atoms, **will be published elsewhere (14)**.

We are much indebted to Dr. Jerry Donohue for constant advice and criticism, especially on interatomic distances. **We have also been stimulated by a knowledge of the general nature of the unpublished experimental results and ideas of Dr. M. H. F. Wilkins, Dr. R. E. Franklin and their co-workers at King's College, London (15)**. One of us (J. D. W.) has been aided by a fellowship from the National Foundation for Infantile Paralysis.

equipment, and to Dr. G. E. R. Deacon and the captain and officers of R.R.S. *Discovery II* for their part in making the observations.

[1] Young, F. B., Gerrard, H., and Jevons, W., *Phil. Mag.*, **40**, 149 (1920).
[2] Longuet-Higgins, M. S., *Mon. Not. Roy. Astro. Soc., Geophys. Supp.*, **5**, 285 (1949).
[3] Von Arx, W. S., *Woods Hole Papers in Phys. Ocearog. Meteor.*, **11** (3) (1950).
[4] Ekman, V. W., *Arkiv. Mat. Astron. Fysik.* (Stockholm), **2** (11) (1905).

MOLECULAR STRUCTURE OF NUCLEIC ACIDS

A Structure for Deoxyribose Nucleic Acid

WE wish to suggest a structure for the salt of deoxyribose nucleic acid (D.N.A.). This structure has novel features which are of considerable biological interest.

A structure for nucleic acid has already been proposed by Pauling and Corey[1]. They kindly made their manuscript available to us in advance of publication. Their model consists of three intertwined chains, with the phosphates near the fibre axis, and the bases on the outside. In our opinion, this structure is unsatisfactory for two reasons: (1) We believe that the material which gives the X-ray diagrams is the salt, not the free acid. Without the acidic hydrogen atoms it is not clear what forces would hold the structure together, especially as the negatively charged phosphates near the axis will repel each other. (2) Some of the van der Waals distances appear to be too small.

Another three-chain structure has also been suggested by Fraser (in the press). In his model the phosphates are on the outside and the bases on the inside, linked together by hydrogen bonds. This structure as described is rather ill-defined, and for this reason we shall not comment on it.

We wish to put forward a radically different structure for the salt of deoxyribose nucleic acid. This structure has two helical chains each coiled round the same axis (see diagram). We have made the usual chemical assumptions, namely, that each chain consists of phosphate diester groups joining β-D-deoxyribofuranose residues with 3′,5′ linkages. The two chains (but not their bases) are related by a dyad perpendicular to the fibre axis. Both chains follow right-handed helices, but owing to the dyad the sequences of the atoms in the two chains run in opposite directions. Each chain loosely resembles Furberg's[2] model No. 1; that is, the bases are on the inside of the helix and the phosphates on the outside. The configuration of the sugar and the atoms near it is close to Furberg's 'standard configuration', the sugar being roughly perpendicular to the attached base. There is a residue on each chain every 3·4 A. in the z-direction. We have assumed an angle of 36° between adjacent residues in the same chain, so that the structure repeats after 10 residues on each chain, that is, after 34 A. The distance of a phosphorus atom from the fibre axis is 10 A. As the phosphates are on the outside, cations have easy access to them.

The structure is an open one, and its water content is rather high. At lower water contents we would expect the bases to tilt so that the structure could become more compact.

The novel feature of the structure is the manner in which the two chains are held together by the purine and pyrimidine bases. The planes of the bases are perpendicular to the fibre axis. They are joined together in pairs, a single base from one chain being hydrogen-bonded to a single base from the other chain, so that the two lie side by side with identical z-co-ordinates. One of the pair must be a purine and the other a pyrimidine for bonding to occur. The hydrogen bonds are made as follows: purine position 1 to pyrimidine position 1; purine position 6 to pyrimidine position 6.

If it is assumed that the bases only occur in the structure in the most plausible tautomeric forms (that is, with the keto rather than the enol configurations) it is found that only specific pairs of bases can bond together. These pairs are: adenine (purine) with thymine (pyrimidine), and guanine (purine) with cytosine (pyrimidine).

In other words, if an adenine forms one member of a pair, on either chain, then on these assumptions the other member must be thymine; similarly for guanine and cytosine. The sequence of bases on a single chain does not appear to be restricted in any way. However, if only specific pairs of bases can be formed, it follows that if the sequence of bases on one chain is given, then the sequence on the other chain is automatically determined.

It has been found experimentally[3,4] that the ratio of the amounts of adenine to thymine, and the ratio of guanine to cytosine, are always very close to unity for deoxyribose nucleic acid.

It is probably impossible to build this structure with a ribose sugar in place of the deoxyribose, as the extra oxygen atom would make too close a van der Waals contact.

The previously published X-ray data[5,6] on deoxyribose nucleic acid are insufficient for a rigorous test of our structure. So far as we can tell, it is roughly compatible with the experimental data, but it must be regarded as unproved until it has been checked against more exact results. Some of these are given in the following communications. We were not aware of the details of the results presented there when we devised our structure, which rests mainly though not entirely on published experimental data and stereochemical arguments.

It has not escaped our notice that the specific pairing we have postulated immediately suggests a possible copying mechanism for the genetic material.

Full details of the structure, including the conditions assumed in building it, together with a set of co-ordinates for the atoms, will be published elsewhere.

We are much indebted to Dr. Jerry Donohue for constant advice and criticism, especially on interatomic distances. We have also been stimulated by a knowledge of the general nature of the unpublished experimental results and ideas of Dr. M. H. F. Wilkins, Dr. R. E. Franklin and their co-workers at

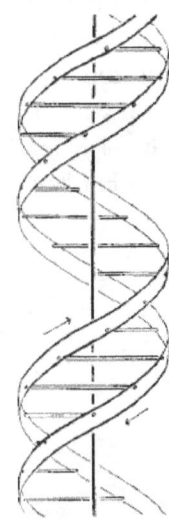

This figure is purely diagrammatic. The two ribbons symbolize the two phosphate—sugar chains, and the horizontal rods the pairs of bases holding the chains together. The vertical line marks the fibre axis

Francis H. C. Crick

When Francis Crick (1916 - 2004) was growing up in England, he received a children's encyclopedia from his parents, which exposed him to the world of science. His fascination with this world has continued throughout his life. He received his college degree in physics and was starting graduate school when the World War II began. During the war, Crick worked on weapons for the British Admiralty. He was in his late 20s by the time the war ended, but he decided to go back to school for a PhD. Around the same time, he read a book that inspired him to begin studying biology. He went to the Cavendish Laboratory of Cambridge University to pursue this interest by studying proteins. In 1951, James Watson arrived at Cavendish, and the two began the collaboration that would lead to the discovery of the structure of the DNA molecule. Before Crick received his PhD, he completed the work that would earn him a Nobel Prize. Since 1976, Crick has been at the Salk Institute in California, where he investigates topics such as the origin of life and consciousness.

James D. Watson

As a boy, James Watson was already very interested in science, particularly in birds. Watson's interest in DNA grew out of a desire, first picked up as a senior in college, to learn about the gene. By the time he got into graduate school at Indiana University, he decided that if he was going to understand genes, he needed to understand the simplest form of life bacteria. He then headed off to Europe, as a postdoctoral fellow, to learn more about biochemistry and bacteriophages. In 1951, he wound up at the Cavendish Laboratory, where he met Francis Crick. In 1953, Watson and Crick sparked a revolution with their discovery of the helical structure of the DNA molecule. Watson was only 25 years old when their findings were published. And he was only 34 when he was awarded the Nobel Prize. Since 1968, Watson has served as director of Cold Spring Harbor Laboratory, a research institute for molecular biology.

From: *accessexcellence.com*

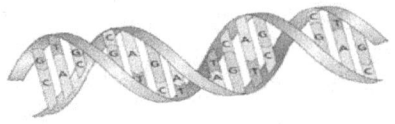

www.ingramcontent.com/pod-product-compliance
Lightning Source LLC
Chambersburg PA
CBHW080842170526
45158CB00009B/2607